How to Talk to Anyone

How Anyone Can Master the Art of Small Talk, Build Stronger Communication and Make a Killer First Impression

Jason Miller

Jason Miller

COPYRIGHT © 2019 BY JSON MILLER

All rights reserved. No part of this book may be reproduced or used in any manner without the written permission of the copyright owner except for the use of quotations in a book review.

Illustrations copyright © 2019 by Ralph Williams

Cover photography by Ralph Williams

First Edition: November 2019

Produced by Jaon Miller

Printed in the United States of America

TABLE OF CONTENTS

PART - I
The Art of Small Talk
How to Master the Unwritten Code of Social Skills

Introduction 6

Chapter 1: Why is it Hard for You to Talk to People? 9

Chapter 2: The Foundation of Social Anxiety and Learning to Cope 16

Chapter 3: Building Confidence for Better Social Interaction 22

Chapter 4: Understanding the Mechanics of Human Interaction 28

Chapter 5: The Art of Small Talk 35

Chapter 6: How to Initiate Non-Verbal Communication 48

Chapter 7: Becoming a Master at Small Talk 57

Chapter 8: Channeling Positivity into Your Conversation to Keep it Going 67

Chapter 9: Using the Art of Storytelling to Drive Conversations 84

Chapter 10: Building Quality Relationships and the Keys to Making Them Last 90

Chapter 11: Social Interactions in a Group Setting 104

Conclusion 110

PART - II

The Art of Analyzing People:

How to Master the Art of Analyzing and Influencing Anyone

Introduction 113

Chapter 1: What's the Problem? – How to Analyze People Instantly Using Proven and Successful Techniques 120

Chapter 2: How Many People Are Gifted with the Talent to Read People Instantly? 128

Chapter 3: Different Types of People and How They Fit in the Social Circle. 138

Chapter 4: Basic but Proven Effective Techniques for Analyzing People 148

Chapter 5: Lies – Why They Affect the Way You Analyze People? 165

Chapter 6: Adverse Effects of Misreading People 176

Chapter 7: Analyzing Verbal Cues 187

Chapter 8: Looking into One's Own Self 195

Conclusion 204

References 210

PART - I

The Art of Small Talk

How to Master the Unwritten Code of Social Skills, Improve Your Charisma, and Little-Known Hacks to Connect with Anyone Effortlessly

Written by
JASON MILLER

Introduction

At some point in our lives, we have all been that socially awkward person. And whether you have come to label yourself as an introvert, or you are an extrovert who is struggling to connect with that inner social butterfly, we have to come to that place where we realize that social interaction with our peers is essential for human growth.

That said, many of us find it difficult to interact with humans. There are a lot of explanations for this, but this time around, I encourage you to stop giving in to those explanations and instead take a stand today to become better at conversations. It is going to be a scary journey, especially for those of us who are shy and reclusive, but that doesn't have to define you going forward. In this book, you can learn how to take charge of your social life and find ways to build relationships that will not only empower and sustain you, but they will nourish you in ways that you didn't think was possible.

Approach this book with an open mind. Let go of any preconceived notions you have about why you are the way you are. As I said earlier, you may have come to put a label over yourself, and what this does is to help you cope better with the distance you have with the people in your life. However, as humans, we are biologically programmed to seek out each other. There is a longing on the inside of you. Something that wants you to reach out and connect. This is perfectly normal. In this book, I share with you my journey to becoming an extrovert. Now, I use the label 'extrovert,' not because I am the typical definition of someone who enjoys being in crowds. I am reflecting on how I left the place where it was difficult for me to even establish eye contact to this point where I can meet a complete stranger, smile and start a conversation. These are milestones that I crossed, and while it was hard at first, I can tell you that over time, it got easier.

If you are planning on becoming a public speaker, this will provide the foundation or premise for that journey. In this book, you will also learn how to interact with people in a public setting because public speaking is more than just standing on stage and talking to the crowd. You have to connect with them, and while it is a lot easier to connect with people when you are in a one-on-one setting, it is not impossible to replicate that effect in a group setting. And that is just one of the many things you are going to learn in this book. So, as you flip to the next chapter, take a deep and positive breath, let go of your fears, and what you think you can or cannot do. Make

Jason Miller

your mind a blank slate because here on out, we are writing new experiences in your social life and we are going to do it in style.

So, turn over and let us begin.

Chapter 1

Why is it Hard for You to Talk to People?

Before we start reeling out a three-point solution to the problem at hand, it is important to get to the root of the problem first of all. And to get to the root, we seek to answer the questions of why we are the way we are. The better understanding you have about why you act the way you do, the more effective you become at implementing the solutions. More than that, knowing the root of the problem gives you deeper insight into your personality and helps you make sense of the world around you.

Why is it Scary to Talk to People?

If we are going to look into all the reasons explaining why talking to some people might seem like a scary experiment, we will be here until 12 Sundays from now, and we still would not have exhausted half of those reasons. To save us from that stress, I will focus on just one reason. And that reason is very simple; we project our perception of ourselves on other people. Let me explain that.

When we try to talk to people (and by 'we' here, I am talking about us shy folks), we think of how they would react towards us. Before we meet people, we have a certain image of ourselves when it comes to how we look and how we sound. And often, we feel that these perceptions do not match up to the required social standards. And so every time we meet new people, we feel that they are judging us based on these opinions we have about ourselves. In other words, we think for people when we come across them. And then, on top of that, we insert our negative opinions about ourselves into those thoughts that we are thinking on behalf of those people.

Do you realize how ridiculous that sounds? But that is essentially what we do, and because we think that people are thinking these things about us, it makes it difficult to approach them. These thoughts are not always self-conceived. As in, we do not just sit down and create those thoughts. More often than not, these are things that have happened based on experience. Probably in our childhood, we had a social setting where we were embarrassed by our peers who brutally capitalized on our insecurities. Or perhaps, our parents may have directly or indirectly sowed the seeds of self-doubt in our hearts and we grew up with this unpleasant notion about ourselves.

Another possible root for the negative opinions we have about ourselves could be from the kinds of content that we feed our minds. There are a lot of books, magazines and content out there in the world telling us how we should look. And when we are different from these things, we develop insecurities about those differences. Our fear of talking to people comes from a place of insecurity and to get over it; we would first need to get over ourselves.

Where Does Fear Come From?

Fear is a biological response to anything that threatens your being. When you are afraid, the fear that you experience triggers a survival instinct bent on preserving you. So, if your fear is activated when you get into a social setting, essentially, your mind or brain is trying to preserve you from any perceived dangers in your environment, and this happens because your brain has come to look at social settings as a place that threatens your well-being. This is not something you were born with. It is something that is emphasized over time, like a habit.

Fear is not the horrible monster we have come to know. We may not be like how it makes us feel, but if you look at fear from an objective perspective and get an understanding of it, you would realize that you can harness it for your own good. Fear keeps you alert. Now, when you indulge in fear for too long, you become paranoid. However, if you are able to plug yourself into the root of your fear, you can use that knowledge to empower yourself and what do I mean by this?

Since we now know that fear in social settings is triggered because your mind has been conditioned to think that you are being threatened every time you get into a social situation, you can now work out a process of reprogramming your mind to think the opposite. This is not something that is going to happen overnight, as it would require deliberate effort on your part. However, with the commitment, you can make it happen.

I should point out here that for a small group of us, the fear that we experience when we get into social situations is part of our biological makeup and I cannot fully go into the details of that here. This because it is something that will require collaboration with your doctor if you fall under that spectrum. That said, there are still tips you can pick up from this book along the way. I believe that biological makeup or not, there is a psychological angle to this and that is what we want to tackle in this book.

Mental Barriers to Talking to People

Have you ever been in a situation where you finally walk up the courage to talk to a person only to find yourself paralyzed completely on the spot? Even after you spent days to rehearse your lines and conversations, the moment it gets to that point where you were supposed to step onto the plate, you lose it and have that deer caught in the light situation. It is not pleasant. I can tell you for sure because I have been in that boat and there are a lot of reasons for it. Most of which have to do with fear, but seeing as we have already talked about fear in the previous segment, let us look at other reasons that explain the embarrassing situation.

1. Insecurities

We touched on this area earlier when we first started out this chapter. And we are discussing this here because this is a major contributory factor to the mental barriers we experience in our attempts at social interaction. When we try to initiate conversations with people, we are often focused on who we think we are. To make matters worse, we have negative opinions about who we are and this acts as a block in our ability to talk with people.

2. Assertion of Assumptions

When we meet a person, based on certain poor analysis like how the person dresses, the way they talk and so on, we judge them. And based on this judgement, we react. What we have done here is basically to assert an assumption that we have about the person. This causes conflict in conversations because we are unable to get past this mental opinion that we have generated about this person. There is a common phrase that says," do not judge a book by its cover." Many of us take a look at the external attributes of a person without really looking inward to know a person. And without that insight, it is going to be impossible to have a genuine conversation with said person.

3. Language

Language is one of the most complex aspects of human communication. And here, I am looking beyond the differences in our mother and focusing on our perception of what is said and being said to us. Words connote different meanings to each of us and for this reason, we develop different reactions to certain words and phrases. This would explain why certain phases that may have caused a certain group of people to laugh might cause another person to cry. These differences in language can make it difficult for us to talk to people, especially when you consider the fact that shy people are typically very sensitive.

Now that we have set up a premise for why we find it difficult to talk to people, the next course of action is to look into the fears that we face every time we get into a social setting and explore effective ways to overcome it.

Chapter 2

The Foundation of Social Anxiety and Learning to Cope

Social anxiety happens when a person develops a fear of being rejected, judged or negatively evaluated and this feeling of fear is usually brought on or triggered by being in a social setting. The manifestation of social anxiety is usually in one's performance, so in a roundabout way, you can say that social anxiety is performance issues triggered by fear from being in a social setting. If any of this sounds like something you can relate with, you are not alone. Millions of people all over the world suffer from some form of social anxiety or the other. In the next few segments, we will explore the topic in detail and come up with tips on how to maintain high performance even when you are under social pressure due to anxiety.

How to Overcome Social Anxiety

The impact that social anxiety has on our lives is tremendous. However, it does not mean that it cannot be overcome. There are measures you can put in place to help you with the daily steps needed to overcome social anxiety and a lot of these steps are things that you can start doing from your home. That said, it is important to know what spectrum you fall under when it comes to diagnosing a social anxiety disorder. It is a well-known fact that people who suffer from severe and extreme cases would need to rely on drugs and therapy to overcome their struggles. Outside that, these next few steps that I am going to talk about are things that are basically doable from where you are right. If you have already spoken to your doctor, go ahead and put the steps to practice. Build on it and through it; you can build your esteem enough to help you overcome social anxiety.

1. Confront the situation that triggers your anxiety head-on

It is human to want to immediately take yourself outside situations that cause you to feel a certain kind of way about yourself. However, if you are serious about overcoming anxiety, it is important that you put yourself out there. Just ensure that you do this in moderate and controlled doses. Don't plan on going from being a couch buddy to a cliff jumping adventurer in minutes.

2. Keep a journal

An emotional journal is there to help you keep track of your feelings. That way, you can narrow down the specific emotions that trigger your anxiety. It could be fear, anger or sometimes it could just be activities that take you down memory lane to a negative experience that you had in the past. Knowing your triggers will help you become better prepared

3. Get physical

Physical exercise has a way of making us feel good and better about our bodies, not to mention the fact that it helps you explore your mental headspace and gets you into a positive mindset faster.

4. Let go of any illusions you have about being perfect

Most people who suffer from social anxiety disorder have a problem with performance in public. And that is because they think that people expect them to be perfect. It is impossible to be perfect. Let go of the desire to get people to see you as perfect. It is okay to be you the way you are.

5. Stay positive

Being positive is an essential part of your journey to overcoming social anxiety. Train yourself to stop wandering to the dark and negative recesses of your mind. Focus instead on those things that make you feel good about you as a person and the life that you live.

Advantages of Learning Techniques and Education on Social Skills

Essentially, social skills help to prep you for easy integration into social settings that you are not familiar with. It could be for business purposes or a simple education on how to do certain things when you are in public with people. These things that I am talking about are basically what people consider as proper social skills and it can range from basic conversation to traditionally acceptable interaction based on a specific geography. It is important to learn the skills so that you do not end up offending people.

Beyond that, being knowledgeable of common social skills makes it easier to communicate, as certain social faux pas that may be regarded as disrespectful can quickly earn you the tag of rude or difficult to associate with. And when you have these kinds of tags in social settings, you might as well be wearing the scarlet letter on your forehead. People tend to avoid those that are wearing such tags.

In life, you never know how far you are going to go. You cannot always judge your progression in life based on the people you are associating with you right now. It is possible that in the course of your business or career progression, you could find yourself in some of the most amazing cities and places in the world. Beyond the language barrier which could prevent communication, there are also other things that make interaction in these kinds of settings, and having good knowledge of the social skills tell are peculiar to that region can help you understand what is acceptable and proper.

The Case for Learning About Confidence and Social Skills Together

Knowing the importance of social skills is one thing. Implementing them is another and this is where confidence comes in. A person who is confident in their ability and personality does not let that get in the way of them using the knowledge that they have gained to their advantage. Whether they are in the boardrooms negotiating the next big deal or they are pitching their brands and ideas to potential investors ... or perhaps it's a child speaking up for the first time. Whatever category you fall under, confidence is the key to executing the social skills you have gained in real life.

One way you can go about boosting your confidence is by getting rid of the notion that you are not good enough. I had a counselor who used to tell me that, "if you don't love yourself, how can you expect other people to love you in the same way." It is important to build self-care routines as well as indulge in treats every now and then to help you climb out of that hole of insecurities and self-doubt. You need to begin to take on activities that affirm the many skills that you have to reveal the true personality within and discover people who share the same interests with you. In the next chapter, we will go into more detail about confidence and how to build it. This segment is meant to set a premise for confidence and how it plays out and social interaction.

The combination of being confident and having an understanding of how people interact in a certain setting has a way of portraying us as the ideal person that people should relate with. When people have this kind of perception about you, your social life will experience a massive boost and in my opinion, this is a win-win situation for everyone involved.

Chapter 3

Building Confidence for Better Social Interaction

Confidence, as we have discussed previously, is an essential ingredient for any social interaction; whether you are talking to one single person or a group of three, the right amount of confidence can set the tone for that relationship. You may have heard people use the term "doormats." This usually refers to people who are unable to assert themselves in their relationships and this inability finds its root cause is a lack of confidence. Without confidence, you will not be able to express yourself articulately. And when you are unable to express yourself articulately, people who you are in a relationship with tend to overlook your needs. So, it is important to build one's confidence.

Where Does Confidence Come From?

To put it in very short and simple terms, confidence comes from your own perception of yourself and abilities. In other words, the image or concept you have about yourself is what feeds your confidence. If you have a very poor opinion of yourself, there is a very strong possibility that you will not be a confident person. Confidence has been linked to self-esteem issues and that explains why people with poor self-esteem tend to have poor confidence.

However, while confidence is an innate ability, I would say that the society around you can influence the level of confidence that you have. So if you have people or are surrounded by people who constantly affirm your negative opinion of yourself, this would reduce your self-esteem and in doing so, reduce your confidence. The opposite happens when you surround yourself with people who affirm the positive opinion that you have about yourself. Your confidence level will soar and you find yourself doing amazing things.

Now, some people find it easier to develop confidence. You would find children who are born confident and then you have kids who find it difficult even to maintain eye contact with their peers. Over time, the society that grooms them (which includes their family, their friends, their network at school and outside the school and so on) can determine how far the confidence level of this child goes. That said, if you are struggling with confidence, the first step to building it recognizes that you are more than the opinion that you have about yourself.

The Basic Foundation for Growing the Confidence You Need

The most common tip you get when it comes to building your confidence would be to 'fake it until you make it.' Now, this can come in handy in certain situations. However, it is important to understand that there is a fine line between confidence and arrogance. And if you are faking it, your confidence can sometimes come off as arrogance. Genuine confidence has very little to do with pride or looking down on other people. As I established earlier, confidence is all about who you feel you are on the inside and not about the people on the outside.

If you are constantly putting yourself up against the people in your environment and using that as a yardstick to measure yourself with, you have crossed the line of being confident and entered into arrogance territory. That said, here are some tips on how to grow your confidence;

1. Avoid negative places and people

Nothing pulls your confidence down faster than surrounding yourself with people who are constantly negative. Take yourself out of that environment and find a space that will nurture you positively.

2. Do not accept other people's negative opinion of you

People's notions about you may have been fed by some error or mistake you made in the past but that does not essentially define you. If they insist on identifying you with that negative trait or history, that is their problem. You, on the other hand, do not need to deal with their inability to move past that point.

3. Find your voice

For me, I think that this is the most critical part of developing your confidence. You need to find what is important to you and learn how to voice those needs. You may not have to put it into words right away, but acknowledging to yourself the things that are important to you is a great start.

4. Have a strong support system

Whether it is your family, mentor, or basically people who you often get a positive vibe off, it is important that you build your own village or community of cheerleaders. These are people who genuinely care about your well-being and see the potential that you have on the inside.

Asserting Your Confidence in Social Situations

Before going further in this book, I would like you to ask yourself the following questions; what does confidence look like for you in a social situation? Are you one of those people who feel that talking down or ensuring that your voice is the loudest makes you the most confident? If so, it is time to change that mentality. Confidence is more than just how you appear, although, that also plays a role. However, I feel that appearance and confidence basically stem from being comfortable in outfits that you wear.

A lot of people tend to focus on what they wear (in terms of the designer brand), as well as social status. They also tend to lean on their wealthy background or rich educational history like a confidence crutch of some sort. All this is well and good, but they don't necessarily help your standing when it comes to confidence in a social setting. They may open doors for you, but they will certainly not keep you in the room. Confidence in the social setting is basically learning to speak and while speaking is an essential part of expressing your confidence, listening is just as important. When you feel to listen to people, communication automatically shuts down.

It is for this reason that I firmly believe that in social settings, confidence is a juggling act between speaking and listening at the right and opportune time. Confidence in communication also extends beyond the verbal aspect. You can use your body language to communicate and confidence can be expressed in your body language. For instance, slouching your shoulders is considered a sign of poor confidence. Also, one's inability to maintain eye contact can be interpreted as poor confidence. And here is my final tip. When next you are given a handshake, ensure that the handshake is firm. That is a nonverbal display of confidence. These tiny attributes are the things that make up what confidence looks like in social interactions.

Chapter 4

Understanding the Mechanics of Human Interaction

Humans are biologically programmed to seek out each other's company. No matter how much of a loner you claim to be, at the end of the day, there is an unspoken need and desire to connect with other people. This desire is the foundation of human interaction. However, there are laws and rules that guide this process. These laws are not what you find in the constitution. What you should expect to find is the fact that there are socially acceptable behaviors that come to play when you are looking at human interaction. Beyond social skills and etiquette, this chapter was going to go in-depth on how this process works and you get some tips on playing this knowledge to your advantage.

The Basic Psychological Principles of Human Interaction

This segment is meant to focus on why we interact with each other as humans. By understanding the purpose of human interaction and the psychology that guides that process, we better equip ourselves with the right tools (or mindset if you prefer) that will get us to where we want socially.

The very first basic psychological principle of human interaction is the fact that every social relationship that we have serves a goal. Now, that goal may have been plotted out consciously and deliberately. Or sometimes, it is just something that we gravitate to naturally because it suits a need. In a situation where social interaction is deliberate, the person who makes a conscious effort to interact with specific people does it usually to meet a transactional need. For instance, the person may feel that by interacting with this particular individual, they may be able to get the opportunities that would lead to their growth in a work environment or their social status.

This is not to say that the person does it with malicious intent. This person simply sees social interaction as a means to an end. Now let us look at the flip side to this. On an unconscious level, you have this person who gravitates towards a person socially to meet some form of emotional need. This is very common with us, especially if we grew up lacking a parental figure or what we term as an ideal role model in our lives. We try to fill up that hole in our lives with the people that we meet and the relationships that we build with them. When we meet someone who we feel matches the profile of our expectations, we use the relationship that we have with them to replace our losses.

Going forward, it is important to understand why you are entering into relationships with people. And while you are on the subject, try to figure out why these people that you meet might be trying to establish relationships with you. Is it for emotional reasons or transactional reasons? The goal does not necessarily have to determine your acceptance of that relationship. It simply helps you understand where you stand with that person and how to proceed in terms of communication and all of the other topics we have addressed in the previous chapters.

Difference in Manipulation Vs Influence Based on True Connection, Intimacy and Serving Others

Given everything we have talked about so far, it is understandable if you start using the words, influence, and manipulations interchangeably. However, these words have two very distinct meanings and their application in relationships have the ability to break or build that relationship. It is therefore important to understand the differences between these two words.

Because, if you take everything that you learn in this book and decide to use it to your advantage without really understanding the concept behind it, you would find that rather than building good and healthy relationships you have set the tone for manipulations and nobody likes to be manipulated.

Manipulations may yield you the results that you want temporarily, but it could go on to destroy whatever future that relationship had. Influence or the other hand, can impact the choices and decisions that the other person in the relationship with you makes. But, if you are coming from a place of genuine connection and intimacy, it has the ability to enhance communication and promote room for healthy growth. Manipulations involve the use of sinister and devious practices to get your way such practices could include lying, blackmail and threats masked as requests. Other forms of manipulations involve degrading a person to a point where they start to question their own opinions as well as the manipulation of elements in the environment to simulate control.

Influence, on the other hand, employs techniques such as concession, which is basically what happens during negotiation. It also uses authority so people in a position of authority can influence the people that they are leading. Another key principle of influence that a lot of us use is the simple act of 'likeness'. If you like a person, there is a very big possibility that you would go the extra mile for that person and a lot of us use this to our advantage. In this situation, the lines between manipulation and Influence might become blurred, but it tilts more towards influence because your emotions towards that person influence your decisions.

Basic Ideas of What People Think in a Conversation and Social Interactions

If you have ever sat down during a conversation and thought to yourself, "what is this person thinking?", you are not alone. Beyond the subject matter of the conversation that we have with people, there is this innate curiosity that makes us wonder what our conversation mates are thinking. While it is impossible to decode those thoughts at the moment (given the fact that we have no mind reader), there are certain things that can clue us into the mindset of the person. I am not going to go into the specifics when it comes to the thoughts because, different strokes for different folks, as they say. However, I am going to go into some of the different things that can influence the thoughts of a person in social settings.

The Environment

The setting where you find yourself in plays a major role in what the person might be thinking about. For instance, if you are in a work environment, the line of thought would have to relate to the performance issues or anxieties about the performance issues. That is not to say that people in these environments do not think about anything outside the place of work. But this is in line with what is generally acceptable.

The Roles that They Play

The roles that people play influences what they think about in conversations or social interactions. For instance or parents would think more in the lines of things that affect the ability to parent as well as what impacts the future of their wards or children. So, if you find yourself in a conversation with a parent to enhance the conversation, you may want to find common ground by looking at their roles as parents. It is common knowledge that many parents are very proud of their children and the moment you strike up a conversation about their kids, it is always difficult to get them to stop (just kidding).

Their Emotional State

Around Valentine's season, cakes, cards, and flower vendors experience a spike in sales and when you look into their customer base, you find out that a lot of people who buy their products are couples or people who are intending to go into relationships. This speaks to the emotional state of people during that season. My point is, when striking up conversations with people, if you decode their emotional state accurately, you may be able to key into their thoughts and establish a connection with them. During Valentine, this is what these vendors tap into to make their sales pitch.

Chapter 5

The Art of Small Talk

I think the biggest mistake a lot of people make is to assume that small talk is basically those pointless words that you use in an attempt to keep the conversation going. They failed to realize that small talk is the actual engine that drives a conversation. It may feel awkward initially, especially if you are not in an informal relationship with that person. However, if you do it right regardless of whatever phase that relationship is in, you can build a foundation that leads to death if it is what you desire. Whatever your relationship goals are, the fact remains that with small talk you can set the tone for the direction in the area of communication.

The Goal of Small Talk in Conversations

There are several reasons people use small talk in conversations and I will be going into some of them. However, just as I said in the introduction, small talk is a very important engine that is useful in driving a conversation. That said, let us look at the goals of small talk and perhaps, this will help you understand just how important it is.

1. It establishes a connection

Small talk typically evolves around relatable topics and with those kinds of topics, you can use it to get a feel of how the person you are communicating with sees the world without really getting in-depth. This is particularly useful if you are not familiar with the person in the first place. With small talk, you can get to know the person without making an awkward situation even more awkward.

2. It is a coping mechanism for people with anxiety

Social anxiety is more common than you think and we know that it happens when you find yourself in a social setting. If you are trying to break out of your shell and get into that place where communicating with people is no longer tedious, one of the first things you would have to learn how to do is small talk. Small talk is an excellent defense mechanism as it helps you participate in the conversation without dreading the outcome of it and this is because the conversation you are having is in shallow waters, so to speak and offers you safety.

3. It opens up a window for dialogue

This comes in very handy if you are at a networking event and you are trying to get to know people. Rather than coming outright and blurting out your credentials without really getting to know the person, inserting a little small talk into the conversation can set the premise for how the rest of your communication with that person will turn out. If you are able to do the small talk right, the person you are having a conversation with would be more open to hearing the rest of what you actually intend to say.

These three that I have just listed here are particularly useful for the objective that we have in terms of mastering the art of communication. The small talk goes beyond having to discuss the weather and in a few short moments, you are going to learn more on the subject. Just bear in mind that regardless of your personality type, small talk is essential for conversations to thrive.

Conversation Flow

If I wanted to give this topic a definition, I would say that conversation flow essentially is the smooth transition from one topic to another. When you are having an extensive conversation with a person, if you stay on one topic for too long, it might get boring. And if the topic is a sensational one, tempers might flare-up. So, it is important that you consciously apply effort in ensuring that the conversation moves freely from one topic to another. Now, in a bid to stir up the conversation flow, it is also important that you don't just jump from one topic to another. Because then, you start to look like you are unsure of what you are doing. In this segment, I am going to guide you on how to establish a healthy conversation flow even with a stranger.

Step one: Create a doorway for the conversation

A doorway is the starting point for any conversation. Whether you are in an informal setting or any place where everyone is professional, you need a doorway to get you into a conversation. Now informal settings, a lot of people tend to go with the introduction route. They start a conversation by introducing themselves. That works too, but the most effective way to get the attention of a person you want to talk to in a formal setting is to bring the focus on them. What do I mean by that? Say you are aware of what the person does for a living, you can latch onto that as the introduction. An example of being, "Hi, I overheard that you are a software engineer. Do you work primarily with websites or applications?".

A question like that will force them to respond but there's also a great chance that they would respond happily because this is a subject that they are interested in. On the other hand, if you find yourself in an informal setting, a great way to stir up a conversation is by asking questions. However, ensure that the question you are asking is not a 'yes or no' kind of question. It should be a question that will cause them to be involved in their response. An example would be, "Hi there, I am very new at all this and you look like someone who is very much at home here. So, I was wondering if you could make a recommendation for me". Again this puts the person at the forefront of the conversation. If you ask politely and keep your demeanor pleasant, you might be able to start up a conversation.

The choice of the way for the conversation should be dependent on the situation you find yourself in. If you are on a date or an interview, or perhaps you are meeting up with investors, the doorway for the conversation must match the situation.

Step two: Initiate small talk

This is the part that some people have trouble with. But if you pay attention closely, I would say that this is the easiest part. The main ingredient for initiating good small talk is listening. Now, if you end up with a person who is not really into conversations, you may have to do more of the talking, but listen to their responses as well because their responses will cue you in on what the next topic of conversation should be. For example, if you meet the person in a museum and you were able to create a doorway for conversation, rather than ask them about when they developed their passion for art, focus on the little details that they offer you. For instance, if they mention the name of a particular artist, let that be a conversation lead. Make a comparison with the artist that they mentioned with another artist and get their opinions on it.

This is small talk, but in a way, it is helping you get to know the person. So, create the doorway then listen to their responses. Pick up on topics that you are knowledgeable about and find that they are also interested in to initiate the next topic of conversation. It is also fine if you allow the other person to lead you into the next topic. Avoid being monotonous in your responses. Giving yes or no answers when you are having a small talk is a big conversation killer. So, endeavor to respond in three to four sentences and perhaps ask a few questions of your own. As you ask your questions, try not to sound as though you are interrogating the person because that makes people defensive. Instead, affirm their choices and if you don't agree with them on certain topics, politely express your view without using it as an opportunity to convert them.

Step three: Time your conversation

If you keep the conversation short and sweet, people are more likely going to remember you. However, if you continue droning on about subjects that you find fascinating, there is a very big possibility that if that same person sees you at another event, they will avoid you. Especially if they don't share the same interests with you. So, when you get into conversations with people it is important that you time it in such a way that you are able to exit the conversation when the excitement is still high. There is a proverb that says, "leave when the applause is at its loudest." This applies to conversations as well. In a bid to keep your conversation short, I do not encourage you to be checking your time because of that in its own way as rude. However, there are body signs given by your partner in conversation that you can use to decide if it is time to exit that conversation.

For example, if you start noticing that the person you are talking to is glancing about the room, that is a sign that they are looking for someone else to talk to. At this point in time, this is your cue to step back. Another important cue to look out for is if they are checking their own time. These subtle body languages are informing you that the conversation has come to an end. However, if you find that the conversation is riveting with both of you being reluctant to end it, I would still say that for a first meeting, especially if you are in a formal setting where the goal is networking, you should try to end the conversation. That said when you end the conversation with someone like this, ensure that you take their contact details as this could lead to more conversations or communications in the future. But as of the moment, your focus in this setting is to network and you want to mingle and meet up with as many people as possible.

If you are going to stick to time, then I will say in a formal setting, 5 minutes is a lot to spend with one person. That should not mean you end the conversation abruptly. You listen to what they have to say, express your fascination with their ideas and then inform them that you were pleased to meet them; however, it is time for you to move on. Exchange cards where possible and exit the conversation politely.

5 Principles For Success In Conversations

Given everything we have learned so far from the art of small talk to ensure that you have a smooth and healthy conversation flow, this segment is going to break down the guiding principles of a successful conversation. This will go on to help you identify those elements that make the conversation interesting, relatable and most importantly guarantees a repeat;

1. You are good with descriptions

In the mouth of a good conversationalist, words are like the paintbrush in the hand of a painter. The words that you use help to paint a mental picture and that picture is what your conversation partner would identify with. Your inability to use words to describe or articulate your thoughts is the reason why people have miscommunications. Because the words paint a different picture from the message that is being passed across. This is something that you are going to have to learn and groom yourself in.

2. Creative use of contrasts and comparisons

When having conversations, the comparisons and contrasts that you use have a way of enriching the image that you create. For example, instead of just saying that the beauty (of a person or thing you are describing) is delicate, you go on to say the beauty of that person is as delicate as a rose flower. What you have done there is to make the picture you are painting richer and all the more interesting. Comparisons that you use enhances the richness of a conversation.

3. The use of body language

The use of body language in a conversation is essential for the success of that conversation. Whether you like it or not, subconsciously, you are sending out messages with your body. Now, if you make a deliberate effort to ensure that the gestures you make and the facial expressions that you have matches the tempo of your conversation, you sound more interesting. And the reason for that is, your body language animates the conversation.

4. Voice inflection

Excitement can be infectious and the reason for this has been linked to the way we express our excitement. Apart from our non-verbal communication, which includes body language and facial expression, there is also our voice inflection. Have you noticed that when you are excited, your voice pitch takes on a different note? The ability to control your voice inflection is the reason why a lot of radio personalities able to drive interesting conversations over the radio even though you are not actively involved in that communication. When your voice pitch takes on a single monotonous pattern, you become bored quickly. That kind of speech pattern is best reserved for bedtime routines for children as it has the ability to induce sleep. During conversations, you want to keep it interesting so, learn to fluctuate in your voice pitch. Just remember that when you go too high, you sound crazy and when you get too low, you sound weird. Keep your pitch range somewhere in the middle.

5. Interesting topics

When you are having conversations with people, the topic focus should not just be a subject that you find interesting. Your partner in the conversation has to find that topic interesting too. That way, both your interests intersect. You may be one of those people who can talk about the incubation period of the caterpillar and find it so fascinating. However, not everybody is interested in the sordid details surrounding this phenomenon. So, as I mentioned in the previous segment, pick up cues as to what both of you find interesting and elaborate on that. If you apply all of the four previously mentioned principles of a successful conversation, it is bound to make the topic even more engaging.

Having laid out these five basic principles, I feel like I should point out here that no conversation can truly become successful if you fail to listen to your conversation partner. The cues for your conversation guide will come from the details obtained when you listen to that person. You may enjoy the sound of your voice and you may have some interesting contributions to make to that conversation; however, you must remember that conversation is not entirely about you and therefore, you need to give room for the other person to share their thoughts and opinions.

Chapter 6

How to Initiate Non-Verbal Communication

Before you open your mouth to speak, there are ways that your body can communicate. The information that you put out there with your body even though it is subconscious could go a long way in helping other people around you form an opinion about you. For instance, if you are in a public space and you are standing with your arms crossed across your chest, people immediately get the way that you do not want to have a conversation. And even though this is not your intention, your body language is saying you are a no-go area. So, this chapter is about helping you learn how to be more deliberate in the way you communicate without saying a word.

How to Smile

A smile is an involuntary response to something that makes you happy or brings you joy. There are a lot of scientific studies that have been conducted on the benefits of a smile and the results of these studies have linked smiles with longevity, health as well as attractiveness. It is said that smiling makes you more physically attractive and not just to yourself but to the people around you. Now, this attraction is not just about physical beauty. What this means here is that a smile can act as a social magnet that attracts people to you. When you smile in a public space, essentially, you are speaking non-verbally and telling the people around you that you are approachable and available for a conversation.

I did say that as well as an involuntary response; however, it doesn't mean that you can only smile when you feel happy or see something that gives you joy. You can deliberately initiate a smile even though you are not experiencing any of these emotions I mentioned. When I was younger, my mother was very fond of saying this, "fake it until you make it," and she was talking about my smile. I wasn't much of the smiler and anytime we went out, she was constantly encouraging me to smile. At first, when I curve my lips upwards in an attempt to smile, I felt ridiculous and somewhat embarrassed. But I found that as the seconds progressed, I genuinely started feeling like I needed to smile. Going by my experience, of course, in that first stage, the smile can seem painful. If you have ever witnessed a painful smile, you know what I mean... that one where your teeth are grinding against each other and the muscles on the side of your face become strained from the effort you are putting into a smile. That, in my opinion, is not a smile. It is a grimace.

The key to faking a smile is keeping it as natural as possible. And since we already know that a smile is an involuntary reaction to something that gives you joy, we start by focusing on elements around you that make you happy. It could be the color used in the decoration; it could be the snacks that were served, or you could focus on people that you actually like. If the atmosphere around you does not appeal to you in any way, then you could focus internally. Draw on images that make you smile; it could be an old joke or the thought of engaging in an activity that you enjoy. Let these pleasant memories motivate your smile. Now when you do smile, try not to display all of your dentition at once because then you look ridiculous and scary. A small but genuine smile is all it takes to get the wagon rolling.

How to use your eyes/body language

In romantic folklore, they say that the eyes are the windows to the soul. I don't know how true that is, but I do know this; if you do not channel your gaze properly, you can come across as one of many things and one of those things could be the vibe that you are an unpleasant person to be around. When you step into a new environment and you find yourself glancing around, people read you as someone who is nervous and with something to hide. If you decide to glare at people every time they look at you, the interpretation they will get is that you are a person with malicious intent. This also happens when it comes to body language. If your arms or legs are crossed, whether you are standing or sitting, it reads as though you are turning people away. The message that they get from you, essentially, is that you are not available for conversation.

Even if you are not having a one-on-one conversation with people, if you find yourself in a position where you are on a stage talking to a crowd of people, your body language can either support the words you are saying or contradict them. You can use your eyes to express interest in a person or express your disgust of their person. Your eyes can also make a person feel intimidated by your presence, or you can help that person feel as though they are welcome into your space. The language of the eyes and the body is one of the most basic forms of communication. When you were born, you did not have the gift of garb, yet somehow, your parents can read your expression and understand that you have certain feelings about certain things… even though you never expressly stated those emotions. As we grow into languages and understand how to communicate without words, our eyes and our bodies still have an essential message to pass across.

More often than not, I believe that our bodies have a way of communicating the true intent of our hearts. Even if your words are saying one thing, people are paying attention to the way your hands are moving and to the expressions on your face. I remember a hilarious incident that happened when I was much younger. My younger sister was being introduced to lime. We talked about the lime; about how sharp and bitter, the taste was. But this particular brother of mine was bragging about how he loved the taste of lime and so, my mother bought lime for him that day. He tasted it and of course, immediately, he was hit with the sharp sensations and although the words that came out from his mouth were, "I love it," we saw his facial expression and it told us everything we needed to know. His eyes were squeezed shot which basically told us that he could not stand the taste and his face was scrunched up in a way that further emphasized what the eyes were already telling us.

Our body language and our expressions during conversation may not be as intense as my brother's lime lying experiment, however, for people who pay attention to your body language, they get the true message just as clearly as we did.

How to Come Across and Credible and Confident Without Words

Even for the most social person, I would say that it is not every time they walk into a room full of people that they immediately feel confident. They are simply just better adept at masking their emotions. In this segment, we are going to focus on those non-verbal ways to communicate your confidence even before you say a single word. There are several books that focus on this subject because of the volume of information available on this. However, I am going to keep it short and simple by focusing on the basics. As you continue to practice your conversation skills, you would evolve and grow this list.

1. Dress the part

The dressing has a way of building your confidence, and when you look confident, you feel confident. If you are invited to an event, endeavor to put work towards ensuring that the clothes you wear to the event match the theme. That way, you do not stand out as the oddball. One thing I would like you to take note of here is that dressing the part does not mean you have to spend a lot of money on your clothing or wear designer gear. There are three things you need to focus on when it comes to dressing the part. One is your comfort, two is your style and then three is the fit of the outfit.

Comfort does not necessarily mean you have to wear slacks and bunny slippers. Comfort is ensuring that you are able to move around comfortably in that outfit without feeling as though your dressing is going to fall apart. It also means prioritizing what makes you feel comfortable over what is fashionable. So, a nice pair of flats may be more comfortable for you than a six-inch heel. The heels are very attractive and fashionable, but you risk falling down and injuring yourself. It is better to stick with sensible shoes. Choose a pair that are formal. The next part is your style. If you are not into high fashion drama, there is no need to get into the latest trends. Stick to what you genuinely enjoy. When it comes to style, again, all you need to keep in mind is ensuring that your personal style is in tandem with the theme for the event you are attending.

Finally, ensure that the clothes you are wearing are fitted. When clothes match your body size perfectly, it enhances your strengths and hides your flaws, making you look exceptionally beautiful or handsome, as the case may be. This goes on to provide you with confidence.

2. Maintain a good posture

Slouched shoulders, hunched back and a low-hanging chin are classic indicators of poor confidence and low self-esteem issues. So it is important that you maintain a good posture. Stand upright, keep your head held up high and avoid slouching your shoulders. The key to attaining the perfect posture is just like attaining a natural smile. You have to try to keep your posture as natural as possible so that you don't come off looking stiff. Because, if you look stiff, the message you are passing across to people who might be coming your way is that you are someone who is not fun and we both know that you are an amazing person for people to know. All you have to do is give them a chance. It starts by keeping the doors open. What do I mean? Ensure that your posture says you are welcoming, inviting and accessible.

3. Don't run away from direct contact

When you meet someone, you instinctively reach out your hand for a handshake. This is a ritual in communication that must be adhered to. When you are in a handshake, ensure that your grip is firm. However, don't make it too firm that you hurt the person. Another thing you have to try to do is to maintain eye contact. Averting your gaze when there is a contact might be interpreted as you being either shy or lacking when it comes to confidence. A confident person is able to hold another person's gaze. If you feel that it is too tedious to maintain eye contact with a person, try this simple trick I learned. Gaze at the person for 10 seconds, then look away to a distance that is no longer than 10 feet away from you for another 10 seconds before returning your gaze to the person. This way, you don't have to feel intimidated by that process. You engage the person with your eyes, take your gaze away for a few seconds and then return the focus to that person. In a more intimate setting, I would advise that you hold that gaze for longer. It builds the connection you have with that person.

Chapter 7
Becoming a Master at Small Talk

Small talk is the engine that drives a conversation which eventually leads to building lasting relationships. If you are able to master the art of small talk, you will become an excellent conversationalist. People love to talk with someone who has something interesting to say. But more importantly, they also want to be heard. Small talk is not just about you talking.

It is about creating an atmosphere that allows for easy communication and communication can only happen if you allow the other person to have room to express their opinions. And just so you know, the subject of the small talk does not really matter as long as you are able to handle it delicately.

You could be discussing the weather and still make it such a fascinating subject that the other person involved becomes engaged. This is what it means to master small talk.

How to Start Small Talk

Knowing what we know now about small talk, here are a few pointers to help you get started;

Step one: Put your phone aside

Small talk is not the time to start showing off the make and model of your phone or trying to view the latest happening on Instagram. Social media has made it possible for us to connect with people all over the world; however, because of the numerous voices on the platform, it has made us more disconnected from our real world. I watched a movie once where this nice family was having dinner. And the whole room was quiet, not because the food was super delicious, but because everyone was on their phone. A platform where the family was supposed to connect with each other gave room for a total disconnect because of phones. So it is important that you disconnect from social media and your phone in order to connect with the person in front of you.

Step two: Ask open-ended questions

I talked about this earlier, but I didn't go into detail because I knew that we were going to run into this here. The questions that you ask will elicit a response from the person that you are talking to. If you ask simple 'yes or no' type of questions, that is exactly what you would get. You need to learn to ask the type of questions that would require the person responding to use more than two sentences.

Step three: Be enthusiastic

Remember what I said about voice inflections earlier. When you inject excitement into your words, it automatically triggers excitement in the other person. This is because, as humans, we are empathetic by nature and one of the ways we show empathy is by mirroring the action or response of the other person in conversation. Your enthusiasm can become contagious simply because the person empathizes with where you are coming from and they are now mirroring your reaction. It may feel a little fake initially, but the more you practice, the better you become at being excited about the topic of your small talk

Step four: Listen attentively

One thing I have always said in this book is that the foundation for your small talk is basically in the answers that you receive. When you ask those open-ended questions, the responses that you get are what will build the topic for the next set of questions and that is essentially how your conversation will grow. So, pay attention. Besides, you never can tell when the other person would ask a question of their own and if you weren't paying attention or listening to anything they were saying, you might end up making yourself and the person look like a fool.

Step five: Choose your small talk topics carefully

Small talk is not a time to express your political and religious views. Not only is there a very strong possibility that you might end up offending the person you are having a conversation with, but it also creates room for animosity and can make the person that you are having this conversation with feel defensive. When people are defensive, their guards are thrown up, leaving you on the outside. Safe topics for conversation include arts, sports, hobbies, professional interests and of course, my personal favorite, climate.

How To use the FORD Method For Small Talk

The last segment rounded off with the instruction to be careful about the choice of topics for your small talk. There I listed a few examples; however, the Ford method is one of the most reliable ways to decide on the preferred topic for small talk, especially if you are having a conversation with a complete stranger. The Ford method essentially focuses on the acronym FORD, and it means;

F = Family
O = Occupation
R = Recreation
D = Dreams

These are the safest topics to go for when initiating small talk with new people. One thing you should bear in mind, though, is the fact that people react differently to certain questions. This is because they all have their own personal experiences and those experiences are not always pleasant. For instance, a person who is not in very good standing with their family may not be too excited about talking on that subject. It is now left for you to take cues from what they say as well as their body language. The second you get that standoffish or defensive vibe from them, that is your cue to drop the subject. Do not proceed if you sense that the person you are trying to have a conversation with is not comfortable with the choice of topic. That will only ruin things for you and them.

How to Ask Excellent Questions

You would not believe the amount of diplomacy that is required in sustaining a healthy conversation even if you are just meeting that person for the first time. The reason for this diplomacy is that we live in a very sensitive world and people are now beginning to discover and leave their personal truths. Your opinions no longer govern the lives of other people. Now, I am not saying that it did back in the day. It's just that people were more tolerant of outside opinions. Not to mention the fact that they were very into ensuring that their lives were pleasing to those around them.

But all of that is rapidly changing and for this reason, you have to apply caution when you ask certain questions. Even if you are genuinely curious about what is going on in that person's life and you are certain that you have no ulterior motive, you need to respect the boundaries that people have put up and unless you are invited, certain doors on certain subjects will always remain shut. This segment is about helping you find a way to ask those open-ended questions that will not offend or cause the person you are asking those questions to feel insulted. See what I mean about being diplomatic? Anyways, let us get into it.

As I mentioned earlier and open-ended question is basically the kind of question that would require the person in the conversation that has been asked to respond in more than one sentence because it would require them to use their own knowledge and feelings an example of an open-ended question is, " what have you been up to today?". You see that there is no way a person can answer yes or no to that question, they would have to break down how their day went even if they may prefer to give the paraphrased version.

Typically, open-ended questions begin with any of the following words; why, what, describe and explain. The last two words on the list are not really questionnaires on their own; however, they can be used to elicit a lengthy response from the person you are conversing with. If your questions begin with a will, do, are, and so on, the response you are going to get is going to either be a yes or a no with little to no explanation.

The Power of Listening

There are several definitions of the word, listening. However, I would say that the one that most appeals to me is this one that says, "listening is the mindful act of hearing and making an attempt to comprehend what is being said by the other person." In other words, listening is not just hearing the words that being said to you. It is a choice that you have to make to understand what is being said to you. These days a lot of people feel that communication means talking and because we all have something to say, we feel that it must be said. And in our bid to ensure that what we have to say comes out, we fail to pay attention to the words that are coming out from the mouths of the people around us.

Active listening I am told is an essential communication skill. You cannot call yourself a great conversationalist if you fail in the area of listening. What that says about your person is that you are only concerned about your voice as well as the voice in your head. Every other person's opinion might as well be dust;

you receive it, but you never make use of it. No matter how smooth you are when it comes to the gift of garb, if you are not a listener, you will not be able to engage your audience. Even in shows where you have comedians on stage doing their bit, despite the fact that the only person doing the talking is the comedian, he or she still listens to the audience for cues and it is those cues that help him deliver his performance exquisitely well.

We have already established that there is both verbal and nonverbal communication. For conversations to go smoothly, you have to pay attention to both what is being said and the body language you are getting from the person. In a situation where you have multiple people talking at once, there is very little possibility of the cause of the problem being resolved and that's because everyone is talking at the same time. The solution will only arise when one person decides to listen and hear the other party out. Mastering the art of small talk and knowing when to listen are two of the basic skills you need in order to become an excellent conversationalist.

Chapter 8

Channeling Positivity into Your Conversation to Keep it Going

Before we go any further into this chapter, I want to say congratulations. You have come a long way from where we started and while we still have some ways to go, it is important to sit back and take note of the progress you have made so far. I also hope that at the end of each chapter, you give yourself social assignments to carry out so that you can experiment on the lessons you have learned. This is not the kind of book that you open and read from chapter to chapter, back to back, until the end.

I think that would be boring and at the same time, it will not help you assimilate the information that I have compressed into this book. So, in case you haven't been doing this already, at the end of each chapter (if possible at the end of each segment), close the book and carry out a physical social experiment. Create journals to record your experiences and proffer solutions with a focus on how you think you can improve in those areas you feel you fell short on. Now that we are done with that, let's get right on to the gist of this chapter.

How to Talk and Banter

On the surface level, talk and banter have the same meaning. However, when you go deeper into the definition of things, especially when it comes to conversations, there is a lot of difference. Talk is all we have been doing since we started this journey together from the first chapter. We have been looking at how to talk better, how to engage in small talk and generally just how to be better at talking with people. Banter, on the other hand, is a playful type of conversation. This is the kind of conversation you have with a person whom you are typically attracted to. Sometimes, we have playful conversations with people who are just friends, family members or even people of the same sex. However, for this segment, we are focusing on playful banter with the opposite sex because yes, that is part of communication too.

One major rule of banter conversation is ensuring that both you and the person (s) engaged in the conversation are aware of the fact that everything said in that conversation cannot be taken seriously. You need to apply the law of taking things with the proverbial pinch of salt because everything that you say in that conversation should not be given any kind of levity. The main ingredient for a successful banter conversation is a great sense of humor as well as this mutual understanding that I just talked about. You should be able to roll with the punches and dish out just as much as you are receiving. I have heard of banter conversations that took on a nasty turn and then again, that is not why we are here. We are looking at how to keep the conversation playful and sexy when dealing with the opposite sex without turning ourselves into maniacs.

Another important thing to remember is that banter is a playful exchange and the goal is to tease each other. This is where you display your wit. And as far as I am concerned, there is nothing sexier than wit and humor. In the spirit of keeping things playful and light, your body language should also reflect this. That stern and stiff body posture are not going to work. You need to loosen up a little. That is not to say that you should hunch your back and drop your chin. We talked about this earlier. You have to look confident. But in this case, you also have to look relaxed. Looking stiff and holding your breath at the same time will make you look constipated and I have never known anyone to like that. Finally, you have to learn to have fun with this. You are not writing an exam that your life depends on. You are simply bantering. Also, the only way to become good at bantering is practice. The more you engage in it, the faster you develop the ability to think on your feet.

How to Always have something interesting to Say

Chances are, you have met that one person who is just such a delight to talk to. They always have something to say that stirs up the conversation. That person is what we call a great conversationalist. The thing is, a great conversationalist is not someone who always has something to say. It goes beyond that. After all, is said and done, the best thing about a great conversationalist is the fact that they enter into conversations without any expectations. They don't even try to control the conversation; they just go with the flow. The only time you would find them actively guiding the conversation is when they feel that the topic is veering off into murky waters. Other than that, they just go with it. No matter what is being discussed, they seem to have an idea about it and have the ability to make a valuable contribution to that conversation. This could also be you and here's how to ensure that you always have something interesting to say;

1. Stay up to date on all current events

If you were thinking that the 10 o'clock news was for older people, I am hoping that after this segment, you will change that mindset. The local news basically, is information about what is going on in the world around us. You cannot be so consumed with your life that you fail to be aware of what this is going on. Your awareness of the latest happenings will form a major part of the contribution that you make in any conversation. A great conversationalist always stays updated.

2. Read a book

This book you have in your hand right now will provide a wealth of information that goes beyond just what you were hoping you will get from it when you opened the first page. This is what all books are like. They open you up to a world that you are not familiar with. They transport you to places and times. You can go back in time to visit the past or take a trip to the future. There is just so much information available in a book. Since information is essential for a good conversation, you may need to develop the habit of reading.

3. Think carefully about your answers before you respond

Unlike playful banter, where you allow the first thing that pops into your head to come out of your mouth, this time around you need to think carefully before you respond. At the same time, do not try to force yourself to sound interesting. People can see through your attempts and they would see as someone who is not genuine. In a bid to make yourself sound interesting, you might end up losing the interest of the person you are having a conversation with. Be relaxed, think your answers through and respond in a way that is respectful and befitting of the question you are trying to answer or contribution you are trying to make to that conversation.

How to Resuscitate a Dying Conversation

There are times when no matter how much effort you put into a conversation, we find that things are slowly drifting into a deep end. It is like watching a drunk man trying to walk on a straight line. He has very little control over his limbs and constantly teeters over an imaginary edge. In the same way, you are feeling powerless to stop the death of the conversation. I have been in that situation so many times that I can read the signs of the back of my palm. The banter slowly dies and then people start fidgeting and avoiding each other's gaze. Occasionally, you will hear someone cough here and there. When you see all these signs, know that that conversation is dying. However, it doesn't have to. By injecting your personality into the conversation and taking control of the wheel without necessarily trying to control what people say, you can get the conversation back on track. Here are my tips to help you do so without losing a single drop of sweat

Step one: Don't take it too personally

A lot of times, we assume responsibility for a dying conversation. We feel that it is because we don't have anything to contribute, or we have just too shy to say anything of value that people will find interesting. But this is not the case. Sometimes, people are genuinely tired and when they are tired, a conversation is the last thing on their mind. You have to respect that and don't try to carry the weight of the conversation. If you get the sense that people are tired, let it go. There will always be other times to chat. This is the one time you should accept that it is not your fault. It takes two to dialogue. As long as you tried to put in some effort and you are not getting feedback, you are good.

Step two: Put your newly developed small talk skill to work

If you have been practicing how to engage people with small talk, this is the best time to put it to work. Perhaps the weight of the previous conversation is making people feel uncomfortable. Small talk eases them out of that discomfort and can get the conversation going again. Remember, we talked about the FORD method. This will also be a good place to apply it

Step three: Ask questions and listen attentively to the response

Remember we talked about asking open-ended questions. This is also a great place to put that to work. The questions that we ask will give you more insight into the person and also give them the opportunity to talk (that is if they are interested in talking). I know that previously, I also emphasized the fact that it is the information that you get during this exchange, that will build on the next question. And this is just how you keep going and growing that conversation

Step four: Know when to call it a day

Here's the thing; at some point, you are going to have just to accept that perhaps this person or persons are not interested in having a conversation. And that is okay. It does not reflect directly on your person. We just talked about this in step one; this has to do with the other person. You can't control how they feel about the conversation or their decision to not contribute to the conversation. This is on them and if you are getting this kind of vibe, I would say it is best to check out politely.

How to Come Across as a Positive Person

People naturally gravitate towards someone that they perceive has warm and positive energy. Nobody wants to stand or talk with the bitter crow. And whether you like it or not, there are subconscious vibes that you give off that tells the state of your mind. The subconscious vibe I am talking about here goes beyond your body language. A positive person has a cheery disposition. I would like to point out here that being shy has very little to do with your disposition. I mention this because a lot of people tend to confuse the aversion that shy people have towards social settings with the aversion that a negative person has towards people.

Even if you are shy, it doesn't mean that you are automatically a negative person. A shy person essentially has trouble connecting with people in a social setting and a negative person on the other hand, also experiences trouble connecting people in social settings. But their experiences are not because of the same reasons.

For a negative person, it has more to do with the character traits and dark personality, which makes people avoid them. People don't avoid a shy person, as a matter of fact, because of the quiet nature of the shy person, a lot of people are not even aware of the shy person's presence. But I can tell you that they see a negative person clearly and make a choice to avoid them completely.

To project yourself as a positive person, you would have to do the opposite of everything that the negative person does

1. Do not look down condescendingly on other people

Negative people have a very condescending manner about them. It is almost as if they feel that everyone that they come in contact with is meant to serve them. A positive person, on the other hand, looks at everyone they meet as equals. It doesn't matter your gender, appearance or social status; a positive person automatically finds a way to connect with you. On the other hand, a negative person is looking for reasons to disconnect from you.

2. Do not talk people down when you meet them

This is a classic negative person move. It is all part of the condescending strategy. Sometimes, their talking down on people is not necessarily because they are very mean. They use it to mask their own insecurities. So, if they find that a person is making a more valuable contribution to a conversation than they are, their strategy is to find a way to break that person's confidence. And they do this by talking down on them. A negative person does not appreciate the confidence in a person; instead, they see it as a threat or a challenge. A positive person, on the other hand, is constantly seeking to cheer you on. So, even when people falter in conversations, a positive person does not take that as a limitation or nuisance; instead, they encourage the person.

3. Do not engage any negative thoughts

If you are in a social setting, it is possible that the ambiance may not be as beautiful or as up to standard as you would like it to be. And these things could be in your head as you are ruminating on the entire situation. However, a positive person does their best to ensure that what is in their head, stays in their head. As a matter of fact, they go a step further by focusing on the details that are actually nice. That way, the thoughts invoke pleasant feelings and these pleasant feelings radiate into the aura. A negative person, on the other hand, takes delight in tearing apart the efforts of other people. And so they have no trouble thinking negative things and then even going a step further to air out their negative views regardless of who they hurt along the way.

Avoiding Excessive Negativity in Your Social Interactions

Now that you have figured out how to project yourself as a positive person, it is time to look at a scenario where you have other people projecting negativity on you. Just because you are trying to maintain a dialogue and build a relationship doesn't mean that you should have to take on their negativity as well. If you have learned anything along the way, I hope it includes the fact that emotions are contagious. Just as people can contact excitement from your own excitement, you can also catch negative emotions from the negativity of other people. Besides avoiding the concept or the idea of being around negative people, there is also the fact that negativity is an energy-draining exercise. If you have ever been near a with a negative person, you know exactly what I am talking about.

You spend resources, energy and effort trying to fight off that negative spirit and by the time you are done with them, you would feel exhausted. And the worst part is that you wouldn't have made any positive progress. So, resist the urge to be the person who changes the negativity of other people. If you find that you are already feeling emotionally drained every time you encounter a particular person, you need to start looking out for your own mental health. Now, stepping out of that situation takes courage. Especially if you are a shy person who is just learning how to connect with people. If you find out this person you have connected with has such negative energy, it is heartbreaking for starters. Secondly, it is difficult to get out of that situation because you know how much you struggled to even get into the relationship in the first place. But that is what the segment is about.

Before we get into how to avoid negative energy, let us talk about the signs that tell you, you are in a negative environment

1. Someone is getting hurt

In relationships, there will be times where we unintentionally hurt the other person. This kind of hurt is never deliberate and the moment it is brought to the attention of the person who is inflicting that hurt, immediately they feel remorseful and take steps to resolve the situation. However, in a negative environment, the goal is to hurt. Whether you are the one being hurt or you and the person you are with are collaborating to hurt other people, that is a negative situation and you need to get out of it.

2. You feel as though you are constantly fighting a battle

Relationships are never easy and that is because you have to diplomatically navigate through your needs, wants and compromises. However, if it feels as though everything in that relationship is an uphill climb, you may want to rethink that situation. Relationships are not really as hard as people say they are. Yes, you will encounter challenges the same when you encounter challenges in life, but if you are constantly fighting and feeling emotionally drained at the end of the day, you need to get out.

3. You have no sense of self-worth

Positive relationships have a way of reinforcing your strengths and helping you work on your weaknesses. If you are in a relationship or a situation where you feel that your weaknesses are constantly on display and your confidence is taking a beating on a daily basis, that relationship is very negative. It lacks the nurturing and encouragement that is abundant in positive relationships.

There are many more signs to look out for but these three are classic. So, I urge you to pay attention to. If you find yourself in that situation, it may be hard at first but you need to look out for yourself and take that bold step to get out of that situation. Now, if you are in a networking event or a social setting that has a semblance of a networking event, there is a chance that you would find negative people. Because where you have people in a gathering, you would find clusters of negative energy around the place. You need to step out of those clusters to avoid negative conversations.

This is how to recognize this kind of situation and avoid them:

1. When people are engaging in gossip or hurtful rumors, it may sound delightful to the ears. But remember, someone is getting hurt in that conversation. Even if the person being talked about is not present, you need to get yourself out of that group.

2. When a group of people seems to have nothing positive to say about the event, you can interject and let them know that you see some positive things. If they ignore the things you have highlighted and still go on to

talk about their negative perspective, you should take that as your cue to leave that conversation.

3. If you find yourself in a cluster where the body language is dark and unwelcoming, this is a sign that the atmosphere there is negative. You don't even need to wait for a conversation, quietly pick up your things and move to somewhere where the ambiance is more receptive.

Chapter 9

Using the Art of Storytelling to Drive Conversations

One of my favorite things about my childhood was the bedtime stories that my parents would read to me before I go to bed. More often than not, they prefer to go outside the books that they bought for me because I was the kind of child who got bored with monotony. So, they had to learn how to tell stories that I had never heard before and I tell you, it was the best thing ever. I think that it was because of those bedtime stories that I learned how to be a good storyteller. Even though I was very shy, for the people that I knew and connected with daily, they enjoyed it when I was telling a story.

It could be something as simple as my experience at the office. I learned how to add little embellishments and use certain words to make my story billboard worthy. In this chapter, I am going to draw from my own experiences as well as the opinions of experts to help you perfect the art of telling a good story. This will go on to help you perfect the art of keeping your conversation partners engaged. It is the art of storytelling that makes it possible for a great conversationalist to talk about the weather as though he or she were reading a transcript from one of Sidney Sheldon's novels.

Principles of a Good Storyteller

For starters, a good story is about an event and how the people in that story reacted to that event. A great storyteller has a good story to tell and so for you to become a great storyteller, you have to start with your story. So, let us explore the elements of a good story

1. It comes from an experience

The one thing you should know is that experiences don't have to be something that you went through firsthand. It could be something that you experience through another person. Perhaps it was a story that they shared with you or it was something that you witnessed. Either way, for a story to be good, you need to involve experience. The reason is that when you experience something, you are able to explore that experience with your senses; you know what you felt, or heard or even smelled

2. It should have a series of unexpected events

The twist and turns in the story are what makes it engaging and riveting to the listener. If your story is something that people have heard consistently, over time, they may get bored or not feel half as connected to the story as they would if everything you are laying out with something that was unexpected.

3. It can be embellished but not fabricated

Even the greatest storytellers and fictional writers draw on the experiences of other people to tell their stories. What they write about is not entirely fabricated. There are elements of truth in their stories. What they do however is to embellish the details and make it as unexpected as possible. For an introverted person, this stage here is going to be a very tough hill to climb. However, with some practice, you will get the hang of it.

How to Keep the Other Person Engaged and Listening

In the last segment, our focus was on the elements of a good story. In this segment, we are going to look at how to tell that story. Now, these are two different things but if you are able to bring them together in harmony, you would greatly improve your storytelling skills. So, let us begin

1. Speak in the language that they understand

Storytelling is lost on your listener if they are unable to comprehend you or the things you are saying. Language here goes beyond the commonly accepted native tongue. It is also about vocabulary. If you are talking with children, you would have to bring yourself down to their level and this would also include using words that they can understand. You cannot bring the University standard language into a preschool class and expect that they would understand what you are saying. In the same way, gauge the audience you are with and measure they are receptiveness to your language and then tone it down or up as the case may be, to suit their level of comprehension.

2. Tell your story from a relatable context

If people can't relate to the story that you are telling, there is a very strong possibility that you would lose them. Your message may be a very powerful and life-changing one but if they cannot understand how it impacts them or where they feature in it, you might lose your audience. One way to make your story relatable is to personalize it. Don't approach it with a futuristic perspective or out of body experience. Bring it down to a level where they feel as though they were a part of that story. Emphasize your personal experiences in that story. Highlight what you felt and how you felt and then tie that into the theme of the present conversation you are having.

3. Be animated in the delivery of your story

We talked about voice pitches and voice inflections earlier on and this goes on to emphasize that. When telling a story, if you use the same monotonous tone and expressionless face in your description, you will lose your audience before you get to the end of that story. Try as much as possible to be animated. Let your voice inject as much emotion as possible into the story. Think of it as a stage performance except that you are not on stage and your audience is not paying for your performance. However, the routine is the same. The main actor here is your voice and if you use your voice inflections correctly, you could create different characters in your story and the audience will connect with each of them.

All of this is something that you would have to learn and practice. Some experts recommend taking Improv classes to improve your storytelling skills. I say that is an excellent idea. Go for it if you can.

Chapter 10

Building Quality Relationships and the Keys to Making Them Last

There is an old African adage that says, "when the handshake extends beyond the elbows, it becomes something else." From the first chapter to the 9th chapter of this book, our focus was establishing the foundation for relationships that first communication you have with people that then goes on to build lasting relationships. Now, we have gone past that proverbial handshake. We are now trying to get to the elbow. This is a different ball game altogether. You would need to learn new skills and this is apart from the ones that you are already developing. Some of it will come naturally to you and some of it will take some patience and understanding to get through it. This chapter is about walking you through that process. I intend to help you build good and healthy relationships no matter how much of an introvert you are.

Connecting with People by Finding Common Ground

Human beings are communal creatures. We are biologically programmed to find and connect with people who we have something in common with. The key to lasting relationships is ensuring that you are connecting with people that you can relate with. If you have nothing in common with this person whatsoever, it can be very difficult for that relationship to thrive. And the reason for this is because at some point it would seem as though the relationship is one-sided. When a relationship becomes one-sided, it becomes a breeding ground for resentment.

Today, the world has experienced the highest rate of divorce than it ever has in recent years. And if you look closely at the people involved and have a conversation with them, you would find that the whole relationship crumbled because of resentment. I have listened to interviews of people who were getting divorced or trying to salvage a damaged relationship, one of the recurrent themes in those conversations was the fact that they felt like they no longer had anything in common with this person. It is very important to have common grounds in a relationship and so before you even get into it, you have to ensure that both parties have something in common.

It could be a business goal if you are talking about formal relationships. Or shared values and belief systems if you are talking about informal relationships. Do not connect with people based on trivial things like their appearance or wealth. Over time, those things will fade away. The important things to look for are their character traits, their personalities, their dreams, and their ambitions. These are things that remain consistent over time. And when you get to know these things, you have to be able to see where you fit in because your fitting into their world as well as their fitting into your world establishes common ground. That way both worlds can come together without colliding. Lasting relationships are built through the merging of worlds and it starts from finding common ground.

How to Make Them Feel You Empathize

In today's world, the concept of empathy has evolved from its original meaning. Without going through the whole psychological babble, let me just put it this way; empathy is basically walking in the other person's shoes. In other words, you put yourself in that person's situation and have a second-hand experience of what they are going through. Today, a lot of people feel that empathy is about saying nice things when people are going through stuff. That is not empathy. That is being kind and showing compassion. Empathy allows you to understand a person's point of view and like I said, for you to be able to do that, you would have to walk in their shoes. In this case, it doesn't have to mean you going through exactly what they are going through. You would have to put your imagination to work. Picture the circumstances that they are going through, insert yourself in that situation and then look at things from that perspective. This will give you a unique understanding of a person's behavior, thoughts, and motives.

In relationships, we tend to focus on our needs and this is because we fail to empathize with those of our partners. When someone reacts to you or something that you did, we take it personally because we feel that it is more about us than them. While I am a strong advocate for making yourself the number one lead character in your life, I feel that if you want your relationship to thrive, you may have to get off your soapbox from time to time and view things from the other person's perspective. When you have the perspective of the other person, you are able to contribute in a more valuable way to that relationship. For example, a couple where one spouse stays at home and the other goes to work every day would have to struggle with resentment if either of the spouses does not take the time to recognize the contributions of the other. And to recognize these contributions of the other person, one person would need to picture themselves in the shoes.

If the partner that does not see how difficult it is to manage the children at home and then keep the home in order, they would fail to appreciate the homeliness that they always come to meet at the end of their workday. And this is because they are more focused on the tedious activities they had to face in the office. This also goes vice versa. For couples in the same situation who empathize with each other, you would find them carrying out activities to make the life of their partners easier. The one who comes home from work would not immediately start requesting things; instead they look for areas where they can contribute to the upkeep of the home. And the same goes for the person who was at home. They would not make demands as soon as the other person comes home. Instead, they will give them space to allow them mentally decompress. This is what it means to empathize with people.

How Listening Can Help You See and Make Connections

When you have an emotional need or concern, you are most likely going to go to a person who you feel would listen to you. This also works with people in a relationship. They tend to air their views to people they feel will sympathize and listen to them. This often comes into play when we are thinking of the male and female dynamics in relationships. Men are not really talkers or great communicators are generally speaking in one on one relationships. They are actionable in nature and tend to be more of doers. Women, on the other hand, invest a lot in communicating their emotions and feelings.

The problem arises when the man feels that every time the woman says something, he has to do something about what she has said. But in reality, more often than not, what the woman needs is someone who would listen to what she is saying and sympathize with her. Listening, in this case, is not just about keeping your mouth shut and leaving your ears open. It goes beyond that. Earlier on, when I was talking about the art of small talk, I did say that one of the things you need to learn is how to actively listen and listening in this context also applies the same way.

You have to pay attention to what the person is saying so that you can hear them and then understand where they are coming from. It also helps you to empathize with the person. If you are able to actively listen to the people you are in a relationship with, you will find that you automatically become their confident. More than that, because you paid attention to what they are saying and you are able to understand and empathize with them, you are put in a better position to proffer solutions to their problems.

Everybody loves to be in a relationship with a problem solver and so it is important that you pay attention to what is being said. In the same vein, do not enter into every conversation with the intention to fix the other person because then every time they have a conversation with you, you would find it difficult to hear them out. This is most likely because you are already playing these scenarios in your head where you are fixing things.

You are not the human equivalent of correction fluid. It is ok if you don't fix a situation. Sometimes, all you need to do is just listen. Everything starts with active listening.

How to make them feel like your family

Family is one of those relationships in life that we have absolutely no control over. You do not choose who your biological relations are. You are basically born into them. However, when you start socializing, you are making a conscious decision to choose who you relate with and sometimes those relations that you build create a bond that can be likened to that of a family. The love, respect and loyalty in that relationship are enough to cement your bond for a long time. But how the people get from the point of being strangers to that place where they feel like this person they are with is a relationship for a lifetime? The main ingredients there is you. As I said earlier, when it comes to families, you have no choice in that matter. But in this situation, you are in complete control.

Without realizing it, we have been given the power to decide how long our relationships last. And before you try to make that argument, yes, it takes two or more to build a relationship. But it also comes from a conscious decision on your part. You would still need to be the one to apply certain principles and boundaries to enable that relationship to thrive. When you put a plant in soil, it is biologically programmed to grow. However, there are things you can do to ensure that the plant experiences excellent growth so that it is able to flourish and bear fruit. This phenomenon also applies to relationships. As humans, we are biologically programmed to connect with each other, but it requires effort on your part to foster and nurture that connection into a long-lasting relationship. When someone becomes your family, you are saying to that person that they are more than their flaws and mistakes. And that they have earned your loyalty and your trust and regardless of what happens in the future, you would always be there for them. There is also the unspoken expectation that they would also accept you in the same way.

I would like to insert here that it is important to ensure that the feeling is mutual. If not, you are setting yourself up for disappointment. Expectations have killed a lot of relationships mostly because we expect things that we can't give or we give things that we don't expect from the other person. There has to be a balance. If the relationship has that mutual feeling, the progression from the level of strangers to the family will take a natural turn. In fact, you probably would not even need to utter those words out loud. It is an unspoken commitment. My advice is this, you have been given an opportunity now to choose who you call family, apply due diligence in that process and most importantly, obey the golden rule of relationships; treat people as you expect to be treated.

The Difference Between Sincere and Fake

Let me start this segment by stating it categorically that if you are faking anything in a relationship, you have laid a foundation for dishonesty. A healthy relationship thrives on sincerity and openness. If there are areas where you have to fake communication and genuine affection for that person, I hate to be the one to break the news to you, but that is not a healthy relationship. If you have ever been at the receiving end of fake sincerity, you would understand how hurtful it is. There are people who manipulate their way into relationships to achieve a certain gain. For them, it is all about their goals and not about the needs of the person that they are in a relationship with. If this is you, I would advise you to desist from that. However, to prevent a situation where you are at the other end, here is how you can spot a fake from the real.

1. It feels too good to be true

Everybody wants to live in a fairytale and a lot of times, because of this idea that we have in our heads, we ignore signs that are glaring at us and focus on those things that we want in that relationship. So, instead of seeing the warning labels, we put on our rose-tinted glasses and ride on the highs that this relationship gives to us. If you find that everything appears to be perfect in your relationship, there is something missing. As I said earlier, relationships are not as complicated as we make it out to be; however, there are bound to be challenged. If yours is free from challenges and it is looking like it is straight out of a fairy tale novel, take a long pause and reevaluate that relationship.

2. You have a sense that something is wrong

Our instincts are designed to alert us to dangers or threats in our environment. In a relationship, if you get the sense that something is wrong, even though you just can't place your finger on it, there is a very strong possibility that something is actually wrong. You can either decide to become a Nancy Drew in that relationship and do some behind the scenes detective work or step back until you are able to figure things out. Either way, something is off and that is your instinct telling you that perhaps your relationship is lacking sincerity you desire.

3. Nobody around you seems to like this person

Now, this is a big red flag. No matter how unlikable a person may be, to an extent, you would still find people who would be rooting for them. However, if you find that the people with whom you share a close relationship within your life are unable to accept this new person, there is a chance that they see something that you cannot. And a lot of times, that thing that they are seeing is the insincerity of this person's actions.

Timing a Sincere Compliment and How to Insert it into the Flow of the Conversation

Complimenting someone in the conversation is a way of affirming your appreciation of the person with whom you are conversing with. However, you have to ensure that the compliment you are giving is appropriate to the circumstance. In a formal setting, for instance, you cannot make compliments about a person's body part. Even if they are about things that are innocent like their Eyes or nose. It would make the person feel uncomfortable and an awkward silence could ensue. Your component should be about a personality trait or their contribution to their profession.

Outside that, the next thing you need to focus on is how you deliver the compliment. You want the high praise that you are offering to the person to come off as natural as possible without seeming as though you have an ulterior motive. In my experience, I believe that the best way to achieve that is to avoid lingering on the compliment for too long, especially if you are in a crowded space. If you have a one on one conversation with the person, of course, you could go on to emphasize that particular compliment. However, you should also note that if that one on one conversation is being carried out in a formal setting, it is best to avoid lingering on that compliment. Simply say what you want to say in the most sincere way devoid of offensive words and then move on.

The art of the compliment (without sucking up to the other person)

For this part, I would say your intentions matter a lot. If you intend to suck up to a person, the compliments that you offer to them will actually come off as that sucking up. Because it would lack the genuineness that is required when offering compliments. Again, you should also avoid lingering on the compliments that you give or constantly repeating them. This is only appropriate when you are in a very close, informal relationship with a person. Perhaps, your friend or your romantic partner. In that scenario, I think you need to linger on your compliments as often and as long as you can.

Another mistake that people make when it comes to paying compliments to people is that they feel if you compliment a person, you cannot find something wrong with them. That is not true. It is possible that you like a person's display of empathy; however, you may have a problem with how intense they are in those moments. And because you care about this person, you call them out on it. In healthy relationships, whether formal or informal, genuine compliments go hand-in-hand with constructive criticisms. If you find that you are unable to criticize a person even though you see glaring things that they are doing wrong, but you constantly find yourself complimenting them, it is safe to say that you are sucking up to them.

Chapter 11
Social Interactions in a Group Setting

Now, we have come to the moment of truth. In the previous chapters, we looked at initiating conversations with people one on one as well as inserting yourself into groups. Now, we are going to take things a step further. We are no longer looking at establishing a relationship with one person. We are looking at how you can carry a group along. And in this context, I am not just going to focus on networking in groups. I would look at you being a public speaker. I know that for an introvert, this is a giant leap, but that is how much faith I have in you. Because if I was able to go from being this shy kid to this person who is always excited to get on stage, I believe that so can you.

Group Conversation Flow

In a group conversation, three things can happen. You are either under the spotlight, which means you are the main person championing the conversation. Or you could be a spectator, which means that you are the one who listens more than you speak. Then you have the last situation, which I feel is the most ideal situation; you are sharing the spotlight with all the people in the group. To ensure that you are not lost in that group conversation, the first thing you need to do is identify what role you are playing.

Are you going to be in the spotlight? If you are under the spotlight, it means that you have to be the one leading the conversation, and by leading the conversation, it does not necessarily mean that you are the only one talking. Remember what I said about great conversationalists. They don't take control of the conversation. Somehow, they are able to immerse themselves in the conversation and still give everyone the room to express themselves. If you would rather play the role of the listener, you need to be an active listener. And that means you pay attention to what is being said. Understand what the person is saying and where they are coming from and then interject occasionally with questions based on the information you have received. Try as much as possible not to go off point

And finally, if you are in the last scenario where everyone is sharing the spotlight, the key to balancing it is to be both the listener and the person under the spotlight. You listen to people talking, get their point of view and then assert yourself in the conversation. There are people who have a tendency to want to talk you down. It is your responsibility to take control of your own voice. Not by increasing the pitch of your voice but by ensuring that you assert yourself. That way, you are making yourself heard and are actively contributing to the flow of the conversation.

How to Join an Existing Group Conversation

Unless you are aiming for drama, I will suggest that you enter the group as quietly as possible. And then before you make a speech, listen to what is being said. Hear the views and opinions of the people that are talking in that group and then based on the information that you get; you can then ask a question. Be warned, though; do not go into the conversation with your guns blazing. That spreads hostility and may make the people in that group feel a little bit resentful of you. Even though your point may be just what they need to get the conversation going, you need to take the gentle approach.

As you continue contributing to that conversation, you can take off your gloves and then get into it really deep. The reason I asked that you delay a little before asserting yourself is so that you are able to get your bearings. Understand what the conversation is about, know where most of the people in that group stand regarding the conversation and then assert yourself to make your own contribution. When you listen, you are able to understand more and when you understand more, you are less likely going to make a fool of yourself. This is after all our biggest fears when we get into conversations that involve groups of people.

Group Conversation Guidelines and Principles for Standing Out and Making a Connection

This here is where you learn the art of public speaking. I understand that right now you are not a public speaker, and you probably have no intention of becoming one. But at some point in your life, you may be called upon to address a group of at least ten people. If you start practicing and preparing yourself for that moment, when that time comes, you will find that you are able to have a conversation with your crowd easily and you would do this without fretting or panicking.

The key thing there is to apply everything you have learned so far. But instead of focusing on just one person, focus on the group in the room. The technique I usually use is to have the mindset that I am talking to one person but act as though I am communicating with the group. So, I try as much as possible to use the space that is available to me. For instance, if you are standing on an elevated podium, staying still in one spot does not really help you engage your crowd. You need to take a few steps at a time and keep your eyes on the crowd. You can choose a certain spot to concentrate your gaze on. Occasionally and from time-to-time, let your gaze go over the people in the other parts of the room.

Be as animated as possible in your conversation. Let your hand gestures be appropriate but do not standstill. And finally, put everything you have learned on the art of storytelling to good use. Be clear in whatever message you are trying to pass across. Let it have its starting point in a story and the story should be something that your audience can relate with. Also remember, language is important. Speak in a language that the crowd will understand and avoid the use of offensive words. For this last part, pretend as though your very existence is dependent on it. Because offending one person in a one on one conversation is bad enough. To do it in a group, that could spell the end of your career and affect your ability to speak in public.

And with that, we have come to the end of this book. But don't close it yet, I still have a few words to share with you.

Conclusion

Once again, I am so proud of how far you have come on this journey, and I am deeply honored that I was involved (even if it was in the tiniest way) to help you overcome your social anxiety and become a better version of yourself. Relationships are important. No matter the lies we tell ourselves because of our hurts and past experiences, we still need people around us. With this in mind, I would like you to subscribe to the knowledge that you are in control of the relationships that you develop in your life. You may not be able to control the actions that people take. However, you have a strong voice in determining the kind of people who stay in your life and those who don't.

Going forward, I want you to develop this mindset as you deal with people. Not everybody has to be your best friend and not everybody has to like you. You are important as a person and you deserve to be treated as though you are important. So, choose people who choose you. Treat people the way you want to be treated. You do not deserve anything less. We may have come to the end of this book but believe me when I say that your journey is just beginning. And if you can step outside of your comfort zone, I can guarantee you that life outside your door is going to be exciting and fun. There is nothing to be scared of.

Even when people tell you 'no,' that word carries a blessing. That 'no' is basically sparing you from the heartache of what the relationship would have been if they said 'yes.' And so, rejection is not the ultimate definition of your life. Some of the people you encounter would choose to walk away from you. That is ok, in my opinion, perfect even. Your mantra going forward should be "***choose people who choose me***." I wish you all the best in your professional and social life. And I hope that from this book, you are able to become better at communication and for that reason, you are able to build better relationships. Please remember to pay it forward by passing the knowledge you have gained here to other people in your shoes.

From me and everyone who contributed to making this book a success, we are sending love, light and laughter to you.

PART - II

The Art of Analyzing People

How to Master the Art of Analyzing and Influencing Anyone with Body Language, Covert NLP, Emotional Intelligence and Ethical Manipulation

Written by
JASON MILLER

Introduction

I had grown up at a farm where my father used to have a herd of cows and buffaloes. We also had a garden of apples that we used to sell in the market to make our ends meet. Since childhood, I knew that I was a bit different because whenever I used to meet people, I immediately formed an assessment of them like what they are thinking and how they are going to talk to me.

One day I was walking down a lonely road along with my friend Jasmine. We had just returned after plucking apples from my garden. They were not for the market but to make pickles at home. Of course, they were not ripe. It was about the afternoon when we were passing across the graveyard. Not much of a haunted place like we see in the movies where the hero along with the heroine are caught by a witch, but enough isolated to send shivers down the spine of every sensible person. When we passed through this place while coming to the garden, the sun was shining bright with full energy, but now it was later afternoon. Also, the sun was nowhere to see as the clouds that seemed to be fragile at noon were slowly covering up the sky. Now they had turned into a thick blanket that wouldn't let a single ray from the sun touch the earth.

As we paced up, a dark and gloomy person appeared to be rising over the roof of a hut that was in the graveyard. I had a hunch that something was wrong. He had not seen us until now but could have if we didn't get off the track and hid behind the bushes. Jasmine wanted to stay until he disappears, but a powerful feeling had already gripped me that we must move on while staying along with the bushes. Of course, this made some noise and movement but I was ready to take the risk. We moved on and once we were past the graveyard, we ran on the way to our home.

When we reached home, we were perspiring and our heart was pounding in our chests. The feeling of how we got away and what would have happened if we stayed there or were seen by that person, would not let me sleep at night. That night it rained like madness. Even lightning struck some trees in the jungle and they were all roasted to the ground. I waited for the morning anxiously. At last, the sun had come out and we were ready to bask in its warmth in our yard. I had taken my breakfast and now I was getting ready to go out to school. It was then that I saw Uncle Tom running toward our house. He reached in a few seconds and broke the news that a lady was murdered in the graveyard by some unknown suspect. It all happened in the late afternoon.

The news crushed me and sent chills down my spine. The horror gripped me so hard that I was unable to speak for at least ten minutes after hearing that. I knew something was wrong with that person. He was giving off such a negative human vibe that I couldn't resist thinking that he was evil personified. Anyway, my hunch and careful reading had saved me. What if he saw us? What if he catches us? Could he have done anything to us to kill evidence for covering up his crime? Could anything have happened?

At that moment I didn't know what reading of people was. I just didn't know how we got away. But I researched the subject and brushed up my skills to be perfect in reading and to analyze people.

What This Book Has to Offer?

This book contains proven methods and techniques that can equip you with the skills to read people in an efficient manner. You can learn the skills and practice them to be an expert on how to judge people. When you have mastered this skill, you will be able to guide your behavior in accordance with how the other person is ready to perceive it. In addition, you will be able to eliminate any kind of misunderstanding that gets nurtured when you misread what other person means. Let's take a look at the chapters in this book.

- The first chapter will define what the problem is with reading people. You will be able to learn how to read people and how to react to them. You will learn the skill of analyzing the head movement and studying the feet movement. You will also get to know how you can avoid manipulation by detecting this kind of behavior earlier on in certain people.

- The second chapter will explain that people are god-gifted with the talent of reading people, but it also explains how you can learn the skill if you don't have it naturally wired in your brain. I have explained some tricks to integrate into your personality so that you can be able to kick off the learning process. Some of these tricks that you will find deeply explained are objectivity, ability to trust your gut and to find out how a person behaves naturally.

- The next chapter will explain in detail different types of people. The most prominent and discussed types include the joker, the loyal and the worker. In addition, I have explained different types of personalities like the observer, the idealist, the adventurer, and the performer. You will learn how a particular kind of person or personality behaves naturally.

- The fourth chapter hits the practical steps to reading people. You will learn how to read body language like the eyes, the hands, legs and arms. Then I will move on to the facial interpretation and analyze what facial expressions and micro-expressions to watch out for when you are reading people. You can take a notebook and write it down for reference when you are still in the learning period for reference. Then comes the turn of the inner instinct of humans and how it helps them walk safely on the road to success. The third section of the chapter will explain the importance of human vibes and how they affect our judgment of others. The human vibe can span around the eye projection, tone of our voice and the physical contact.

- The fifth chapter explains the types of liars and how you can efficiently deal with them. You will learn the techniques to protect yourself from their intentions and also help them mend their ways if possible.
- The next chapter sheds light on the adverse effects of misreading people. A flawed judgment can land our relationship in grave trouble. The chapter explains how we are prone to get confused by mixed signals and how they create misunderstandings. I have stated a number of examples about how a mixed signal can ruin our

relationship. In addition, I have stated the signs to watch out for and how to react when you read a mixed signal. When you have read it, you will be better able to detect and analyze a mixed signal when you are confronted with one, and also act fast to end confusion.

- The second last chapter focuses on reading and analyzing verbal cues when you are talking to someone. It contains examples of verbal cues. Then it moves on from there to explain the difference between verbal and nonverbal cues. The chapter contains examples of a kid and a teacher and how they communicate through verbal and nonverbal signs. You can try it out on your own kids.

- The last chapter explains the importance of reading your own body language and looking into your own self. Unless you are clear about who you are and how you think, you cannot succeed in life. You will learn about the benefits of knowing yourself and concentrating on how your thinking flows. You will learn the importance of asking questions from yourself. You will learn how you can find what you like and what you dislike to make decisions faster. You will be able to know your own body's limits and how you are going to react to certain situations.

When you have read this book, you will feel yourself to be on top of every tricky situation. You will be able to judge people accurately and act accordingly. This book will equip you with proven techniques to analyze people and deal with them. This book is for businessmen, students, job-holders, spouses and almost all other categories of people. You don't need to have any special knowledge before reading this book. Anyone can buy and read this book and be a master of analyzing people.

Chapter 1

What's the Problem? – How to Analyze People Instantly Using Proven and Successful Techniques

By reading people, we don't mean that you have to read their minds like a psychic. Instead, you have to analyze their gestures and expressions to calculate what they actually mean. Reading people is about sensing their intentions like what is running in their heads through their behavior. If you gain this ability, you will be able to ameliorate your intimate and social life. When you have read and understood people, you can easily tailor your way of communication to suit their state of mind. This is how you can make an impact in a conversation.

Read People: Who They Really Are. How to Unmask Someone?

The best way to read people is not to let your emotions get over you. Forget about your past experiences. If you are trying to judge people by your past experiences, you will likely misread them. Pay detailed attention to their dressing. If they are wearing casual dresses like t-shirts and jeans, they like to be comfortable, so if they prefer comfort over hardness, they are unlikely to work hard and grow in a competitive environment. Also, see if they are wearing any pendants or stones. If they do, this indicates their spiritual inclination. This helps you judge in a better way.

Another important thing to take into consideration is a person's posture. A high head posture tells us that the person in question is highly confident. If he or she cowers, they suffer from low esteem.

In addition, the emotions that appear on a person's face tell a lot about it. Deep frown lines on a person's forehead suggest that the person is prone to overthinking. Similarly, if a person has pursed lips, he is most likely in anger and is harboring feelings of contempt. If he is grinding his teeth or has a clenched jaw, this means that he is tense.

Most people don't like to get involved in small talk. It is justified given the magnitude of our daily workload and the preoccupation associated with it. But if you ponder over it for a moment, you will realize that small talk, in fact, offers you a great opportunity to get familiarity with a stranger. You can read how he is going to behave in certain situations. That's how you are able to detect any abnormal behavior.

The Way You Treat or React to Other People Depends on the Way You Analyze Them

Once you have read people, it can greatly help you form your reaction to their questions or behavior. For example, if you have deduced that a person is highly confident and social, you will have to set your tone and posture to match his style. If he is confident but you are cowering, you two cannot have a healthy and productive conversation or collaboration.

Similarly, if a person has pursed lips, she is not in a position to listen to anything productive that you throw in her way because she is perturbed by something and will remain inattentive during a conversation. A person will only attentively listen to what you are saying if you are talking according to his or her mental state. If he is cowering and you are head high, he will feel intimidated by your posture and will not be able to open up his heart in front of you. The conversation is likely to end inconclusively or in a deadlock.

How Can You Be Accurate in Reading Someone Using Human Psychology, Body Language, and Personality Traits?

If you want to read people by means of their body language, you have to take a look at the cues that they share with each other with their gestures. Our face is one of the body parts that have considerable importance. Then comes body proxemics. This includes how your body tends to move in space. The third most important thing is body ornaments like your clothes and the jewelry you wear. Firstly, you need to decode a person's cues like interpreting the information that is hidden in their emotions and personality.

Look to Their Eyes

When it comes to reading other people's language, their eyes can be really helpful. You have to pay attention to their eye-contact and how they tend to look away while talking. If they exhibit the tendency to avoid direct eye contact, this indicates that they are not enjoying your small talk or serious discussion. In addition, this indicates disinterest and also deceit in some cases. You can also sense deceit if a person looks away or to the sides. If the person is looking down instead of looking straight, it means he is nervous. In some cases, it also shows submissiveness.

The blinking rate is also important when it comes to reading people's minds. Blinking rate increases when a person is stressed. When a person is touching his face during blinking, he might be lying to you. If the person is glancing at something, this suggests that he has a deep desire for that very thing. Similarly, glancing at a person suggests that the person desires to meet him or her or wants to talk to him or her. If he is glancing at the door, he desires to leave.

If a person, who you are talking to, is looking to the right and upwards, he might be lying to you. If he is looking to the left and upwards, he is speaking the truth. The reason is that it is natural for people to look to the left and upwards when they are using imagination. (Scott, n.d)

Study the Head Movement

The head movement of the person is also of great importance. If the other person is nodding his head when you are talking to him, it either means their patience or lack of patience. If the frequency of nodding is higher than usual, it is the indication that the person is fed up with listening to your talking and needs respite. If she is tilting her head to the sides, she is interested in your talking, but if the tilting is toward the backside, this indicates that the other person is suspicious or uncertain. (Scott, n.d)

Study the Feet

If a person is careful about his nonverbal signals, there is still one fragile point in which you can study to read what is running inside his brain. Why people miss out on controlling their feet is because they are too much focused on keeping in check their facial expressions and other verbal actions. Naturally, a person points his feet while standing or sitting toward the direction in which he wants to go. If he is pointing his feet toward you, he harbors a favorable opinion of you. If you are in a group discussion and a person whom you are talking to is pointing his feet toward some other person instead of you, this is a fair indication that he wants to talk to that person. One important thing is that feet movement and cues are meant to bypass other nonverbal cues. So, even if his facial expressions and eyes say otherwise, you have to follow the cues you pick by his feet.

How Can You Avoid Manipulation by Reading Someone's Mind?

Manipulators have one objective and that is to achieve their goals at any cost. So, their foremost weapon is using deceptive body language. There are some signs that people use when they are emotionally weak and are talking to stressors. But if this is not the case, they are very likely manipulating you by showing exactly the same signs. They will generally use these gestures to gain sympathy from you. Let's roll on to see what these gestures are and how can you avoid manipulation by reading them accurately.

They Will Rub Their Neck and Hands

When a person is manipulating you, he will rub his hands. This most likely indicates self-serving plotting. On the other hand, if he tends to rub his neck, this also signifies the same thing. The manipulator tries to gain your sympathy through this act.

They Will Stroke Their ARMS

When a person is rubbing or scratching his arms, he might have the full intention of manipulating you. This one is tricky because it is possible that the person has other reasons for scratching his arms such as hives. If scratching of arms comes in combination with neck rubbing, this may very likely be a sign of manipulation.

They Will Tap Their Feet

Manipulators tend to shift and tap their feet. This tapping and frequent shifting of feet indicate that they are impatient or even offended. Their impatience will compel you to make a decision in a rush that may most likely not be in your best interest. (English, 2019)

Chapter 2

How Many People Are Gifted with the Talent to Read People Instantly?

Reading people can be a god gifted ability and you can look for certain signs that show that you have that ability wired in your brain. Upon meeting a person for the first time, you usually have a powerful gut feeling which you just cannot explain in a rational way. You instantly form an opinion whether you like them or not. And, over time, when you get to know their real self, you realize that your gut feeling was right. People cannot always explain how they were able to judge others. It is something in their sub-conscious.

Another feeling that most people have but they cannot express is the power to know other people's thoughts. This also is a natural ability. More than once you might have noticed that you were able to tell what other people had on their minds. For example, you bring up a particular topic and leave your friend wondering because he was thinking about bringing up the same topic under discussion.

Sometimes you can accurately tell if your friend is upset. You don't have to communicate with them to know that. It is just his facial expressions that you have to study in order to reach a conclusion. If you are good at this, you have this talent as a god gift.

Some people really boast of their gut feelings. They are pretty sure of escaping dangerous situations just by following their gut feeling. You might have followed your gut and saved yourself from a dangerous situation. For example, your friends are planning a trip to a lake. You cancel the plan at the nick of the time and later find out that all your friends got injured in a road accident. Have you ever had that feeling?

Some people are naturally blessed with the power to detect if someone is lying to them or not. They can tell if someone is twisting the truth or is modifying it. Perhaps they fabricate a story for their personal gains at the cost of your bencfit but they don't know that you are pretty good at finding out the loopholes in their stories. Their eyes, lips and hands tell you if they are telling the truth or not.

All the above incidents are pretty common to most of us. Everyone has a particular gift to use when he is caught in a difficult situation, but most people are unable to explain its words. It remains in their subconscious throughout their lives. If you are a Sherlock Holmes fan, you can understand exactly what I want to tell you. (How To Read People Like the FBI, 2018.)

Can Anyone Learn How to Analyze People?

The ability to read people is concerned with their gestures and other nonverbal signs coupled with their words. Well, it is a fact that you can have this ability in your genes, but this is something that can be easily learned. You have to memorize different signs to accurately judge what other people have on their minds. This includes studying, memorizing and then using a person's posture, gestures, voice tone, facial expressions and also the willingness for an eye-contact in the middle of conversations. There is no rule to read people because people are different. Some have mastered the art of becoming a conman while others appear to be wearing their hearts on their sleeves. You can easily tell what they are thinking and what will be their next step? (How To Read People Like the FBI, 2018.)

Some Tricks to Learn to Read People

It is impossible that you may understand the exact thoughts of a person, but it is always possible to read how they are acting. With the help of some psychology tricks, you will be able to learn how to read people.

Objectivity

The first lesson for a learner is to be objective. You must not let your emotions and biases cloud your judgment about a person. If you have already put that person in a kind of stereotype box, you are the least likely to be right about them. Think like a neutral person.

Try to Find out the Normal Behavior of That Person

When you are trying to judge a person, you should look out for his or her normal behavior. Sometimes you miss out on correctly judging a person just because the behavior he is exhibiting is his normal behavior. This can be scratching of arms, rubbing his hands or neck, tapping the feet and looking sideways. All these gestures, we have discussed earlier, pertains to some kind of nonverbal cues, but they can be a part of a person's normal behavior. That's why you need to set a baseline for that. Calculate what is normal and what is not. Biting nails can be common with a person and cannot always be a sign of lying. If you overanalyze, you can misread people and damage your relationship with that person. Scrape this gesture and look out for some other sign that crosses the baseline that you have set for that person's behavior. (How To Read People Like the FBI, 2018.)

Are you sensing any kind of inconsistency in that person's normal behavior and body language? If you are finding inconsistency in their behavior, phoneme can greatly help you reach the right conclusion. A phoneme is a basic unit of phonetics. If the person is lying to you or feeling nervous while talking to you, she will have an inconsistent voice tone. The pith in her voice will raise or lower down while she uses particular words. She can overemphasize some words to make you believe them. At this point, you are being manipulated.

You also need to understand the context behind the speech and gestures. For example, if a person is sitting with arms crossed, it can be a sign of unhappiness. But the person's choice of taking up this posture can be due to decreasing temperature in the room. Also, you should take into account the type of furniture the person is sitting on. If the chair has no arms, then the person will naturally cross his arms to rest them. When you are in the learning phase, you should broaden your field of focus. Focusing on just one body sign will land you in confusion, and you will misjudge the other person. (How To Read People Like the FBI, 2018.)

Trust Your Gut

Last but not least is that you need to trust your gut. Pay close heed to make sense of your emotions as well as feelings. You can tell that by studying how you feel when you meet them. (How To Read People Like the FBI, 2018.)

Is It Enough to Depend on Your Instincts When Analyzing People?

Whenever we are caught up in a difficult situation, we are repeatedly told and also we tell ourselves to trust the gut. This can be a bad situation or a bad person who we think is bent on doing harm to us. The gut factor jumps in to take its toll on our nerves. But for some people, gut feeling is so powerful that they think there is no need for reading people and analyzing body language. (Chu, 2017)

A single incident, a new employee, a new boss or a new job send the wheels spinning in our heads as we try to figure out how they will impact our lives. One reason behind this immense popularity of our gut feeling is that it is a simple answer to some complex questions that keep us awake at night. Answer to all questions is: "Trust your gut."

Let me explain and clear the confusion. Our instincts are not like a magic spell. We say a phrase and things start happening or an airy creature shows itself and fills us in about a particular incident or person. Instinctual feelings or intuition are linked to our past experiences and knowledge. The reason why people have different intuitional powers is that their experiences are different. The unconscious part of our brain starts working immediately after we encounter something new. It is like pattern matching. When we see a person smiling in front of us, our brain will match this sight with loads of data that is stored in our subconscious. Then it goes on to draw a conclusion. The process is so fast that the conscious side of the brain is totally unaware of this process. That's how we receive guidance in certain situations when we feel ourselves in danger. From this, we can deduce that if the experiences and knowledge are of greater size, our instincts work better. (Chu, 2017)

It is not enough to rely on your gut to reach a decision on whether the person is good or bad. What if you misjudge a person's intentions? When you realize his real nature, it will be too late. Take the example of a cop who, by nature of his duty, has to make fast decisions. He doesn't have enough time to scan through detailed information before he acts. So, one misjudgment can take the life of an innocent person. So, if we base our decisions only on our gut feeling, we are likely to end up making the wrong decision that could land us in trouble. The verdict is that you have to pair up gut feeling with the knowledge that you have about reading their gestures.

It is not always a good idea to follow your gut. Sometimes it is better to just eliminate the need for your gut feeling. You can do this by befriending someone and talking to them without heeding to your gut feeling. Your judgment ought to be calculated and well measured, and should be free of certain prejudices. Otherwise, you are highly likely to make a mistake and lose a friend. (Chu, 2017)

What Can You Do to Improve That Skill?

You can predict what others are thinking. You can read their minds, theorize what they are thinking and also understand their gestures. From all the data, you can know their intentions and analyze their emotions. On the basis of this knowledge, you can predict what their beliefs are and what is inside their hearts. All this makes your conversation with that person fun and productive. To achieve this feat, you should focus on brushing up your skills of reading people. Let's see how to do that.

- You need to stay focused and also be present in the current moment. Never think about yourself amid the process.

- You must be all ears to others. Listen to them attentively. Read between the lines, try to understand the context of their speech. Also, try to understand what they are not saying and keeping back. First, process their words in your brain and try to deduce their meaning. After that, you can respond.

- While you are communicating, you ought to study his facial features, his dressing, the jewelry and makeup in case of a female. Don't forget to take a look at his or her hair cut. In addition, you should take into consideration the surroundings where you are communicating with that person. To improve your reading skills you have to be efficient when it comes to studying that person's facial features and dressing. You have to invigorate your observation skills. You should not miss out on anything minute to large. Even a slight aberration in the person's hairstyle should click your brain.

- You should not lose focus due to some intrusive thoughts. Keep it on the person in front of you. Detect any nuance in their behavior and follow how it develops.

- To improve the skill of reading people, you need to stay calm. If something is perturbing you and you are not at peace with yourself, you will not be able to read the person in front of you accurately. Inner calm is directly proportional to focus. The higher the level of your inner calm, the greater your focus will be.

- If you are the kind of person who loses patience too soon, you are least likely to read and analyze other people effectively. Sometimes you have to listen to another person's blabber on end in order to get to know them better. Practice it if you lack this ability.

Chapter 3

Discuss the Different Types of People and How They Fit in the Social Circle.

All of us are full of different flaws that make us feel ashamed. We do have strengths that we want to brag about in front of everyone. Some of us prefer to stay natural in their everyday life while others love to take up their favorite persona to get through different hurdles in their lives. Some people like to make their way by deception, lies and manipulation while others prefer to face stumbling blocks but refuse to deviate from the right path. Whatever our choice of being a person in our lives is, the goal mustn't be of hiding our weaknesses as well as dark spots if we have any. We must allow our flaws to be a part of our personality. We should celebrate our flaws. This is what being human is about. When a person takes up a fake persona, he forgets that the people, who are loving him, are actually loving that persona that he has taken up and not that person who is in hiding under the fake personality. The real success is that people start loving us because of what we are and not because of what we are trying to become.

The Joker

The first category is a joker. The foremost feeling on hearing the word joker is of a person who is cracking jokes and laughing his heart out even during sober conversations. Jokers love jokes, costumes and makeup. Each makeover gives them a new look and personality. They love to hide their real looks and nature to others. Generally, jokers are considered harmless but if we bring to mind batman's joker, things get totally different. A scary and nutty person comes to mind who is evil personified. That joker is always bent on inflicting the greatest pain on the people surrounding him. Can you think of a person who fulfills the above personality traits? Do you know anyone who laughs too much, always cracks jokes or tries to tease others while laughing it out? Beware! Jokers are masters of disguise.

The Smart One

Smart people have the ability to mold themselves according to the situation. They learn or are naturally gifted to adapt to changing circumstances. Smart people always remember to read other people's styles to gain more knowledge about them. They tend to see through the motives behind their acts and also their hidden desires to work with them and gain benefits. Smart people are good at conveying their messages through in an effective manner and without making the slightest buzz. They know how to express their feelings in a clear way, which is the most important thing when it comes to building and strengthening a relationship.

Similarly, smart people are very successful in their businesses or jobs. They work hard to learn how to read people and the rest gets automatically easy for you. You can tell if a person is smart by looking at how they behave with you and other people around him. One important point to note is that smart people are very good at taking care of their personal interests, even at the cost of others.

The Worker

Workers are the people who belong to a specific social class that is known for doing jobs for low pay only to live hand to mouth in their lives. The jobs they do low demand skills and labor and also have low literacy requirements. This category of people also lives off on social welfare programs. Working-class people mostly remain preoccupied with their day-to-day expenditures. They don't have time to take up different personas and disguises. Also, they are not smart enough to get a job done in the easiest way possible. Their brains are generally wired to do it the hard way. These people generally wear their hearts on their sleeves. They are easy to predict and are simple to understand.

The Loyal

These people are hard to find but exist. They are reliable as well as truthful. If a person is loyal to you, he shares affection with you and will not leave you when life gets hard for you. Loyal people think from their hearts and always work for the benefit of the people who are close to them. Just like the working class, loyal people are easily predictable and trustworthy.

The Strong

Physically strong people generally have happy temperament. A strong person has higher levels of physical and mental strength. They don't have self-pity; that's why they are confident and good at judging people and dealing with them. Before they judge other people, they try to judge themselves. In addition, they have higher levels of self-restraint. Their nerves are powerful that's why they are patient. They also are good listeners and observers. Their physical and mental strengths make them very good at reading other people and reaching an educated judgment. They don't hesitate to ask for help when they are in need, and also, they are open to helping others.

Different Types of Personalities

People are driven by their nature when they do this or that and leave you wondering why they did something that looked unwanted to you. It is perfectly normal if you think you need to want to understand someone a bit more than you already do. This someone can be a loved one or a person at our workplace. We have to accept the reality that people are not perfect. We are different and it is this difference and diversity that makes this world a colorful and interesting place to live in. When people stay true to their role, they tend to contribute their bit to this diverse world. Just imagine if we were all created in the same way, how the world look would like then. It would be boring.

Take an example of diversity. When a car hits a motorbike in a road accident, a huge number of people gather at the site. Most of them are on-lookers who are just investigating what happened. Some mourn the wounds of the injured while some call the ambulance. Only a handful of them step up and actually help the injured recover their senses. They try to administer to the first aid and take care of them until the ambulance arrives at the site. It is not that those people leap into a house on fire without thinking about their lives. We react differently to different situations. These reactions are triggered by our fears and desires. Sometimes they motivate us while at other times, they just demotivate us.

In analyzing people, you should know the people around you. What they do and how they react to different situations. By knowing their personality types and the fears that guide their behavior, you can improve how you interact with different people. It helps you read people in a more efficient way so that your interaction with them becomes smooth and your analysis of people broadens and deepens. In addition, you can track down your own personality traits as well as faults. Let's roll on and take a look at different types of people in the world.

The Reformer / Idealist

The Reformer is a perfectionist. They have principles and are conscientious. These kinds of people have certain ideals to follow and they come down hard on themselves as well as on other people. They just love to keep them at pretty high standards. They are dedicated and responsible besides having perfect self-discipline.

They are usually successful in life because they tend to get lots of things to happen in a short span of time, and that too in the right way. They are always looking forward to setting themselves on the right path by eliminating their weaknesses. (9 Personality Types – Enneagram Numbers, n.d)

The Performer

As the title suggests, these kinds of people will always be setting goals for themselves. They are highly target-oriented individuals and they believe in doing rather than sitting on the couch and thinking day and night. They are always striving for success. This drive makes them pretty excellent at doing things right. You can find them in a big company, a shop or on the street selling vegetables or fruit. Wherever they are, their eyes are always on the horizon. They have dreams of success and they are in the world to make them happen. These kinds of people are considered as role-models by many other people.

They have their fears that drive them toward the top. What makes them perfect is their urge to become somebody. The fear of dying as nobody makes them state-conscious. Instead of discouraging others, they respect the opinion of other people. (9 Personality Types – Enneagram Numbers, n.d)

The Observer

This kind of people spend time on thinking and are of an introvert type. Their focus always is on gaining knowledge. They also prefer reading their own personality instead of reading others. They remain absorbed in themselves and love to play with different types of concepts. They usually abhor worldly attractions like big mansions, cars and social status. They are always busy in searching for themselves. They prefer to observe what is happening in their brains. You can see that these people will lock themselves in their rooms for hours as they love to understand how things go on. This exclusive behavior allows them to concentrate on what they do, that's why they are usually considered as experts on what they do. As they don't have the social skills that are needed to keep relationships healthy, they get overlooked most of the time.

The Adventurer

These kinds of people are fun-loving people. You will see them engaged in enjoyable pursuits and also, they are often in an upbeat mood. They thrive on pleasure and adventures, which makes them a really positive person. They tend to avoid negativity at all costs, which helps them fight off pessimism and stress really well. They are also very optimistic and don't let tough challenges mar their optimism. They are the ones who always find that silver lining in dark clouds. They stick to that silver lining and turn negative situations really fast and really well. (9 Personality Types – Enneagram Numbers, n.d)

Also, they are highly inconsistent. As they are fun-oriented, they remain in a certain work until the fun factor is alive but shoot out of it once they are bored no matter if the work is complete or not. Completion of projects poses a big challenge to them; that's why they struggle to be successful in the practical world.

The Warrior

As the name suggests, these kinds of people love to throw and take the gauntlet. They are strong and have dominating personalities. You can say they are born leaders and are really confident. They are real alphas. They hate to depend on other people and also don't like to reveal their weaknesses. Instead, they use their strengths to give a cover to those people who are around them as their family and friends. They are always ready to take charge of any situation no matter how thundering and dreadful it is. They love to be the masters of their own fate and they also prefer to take control of people as well as circumstances.

Their inner strength also makes them rigid, straight forward and sometimes haughty and harsh. They cannot tolerate signs of weakness in other people. They are ready to confront others on petty issues. They are always ready to express their anger and frustration on things they don't meet up with their expectations. These are the ones that are quite difficult to understand. Their nature is too intense and volatile to let others read them. (9 Personality Types – Enneagram Numbers, n.d)

Chapter 4

Basic but Proven Effective Techniques for Analyzing People

This chapter will walk you through some basic techniques for analyzing people. You will learn what body signs you have to read in order to understand what is running in the other person's mind. In addition, I will explain in detail the importance of gut feeling and the role it plays when you are trying to read other people. The chapter will also walk you through the importance of emotional energy in reading people.

I have touched upon the topic of studying body signs in the first chapter. This chapter will help you learn in detail what each body sign tells about a person.

Posture

How we carry our bodies speak volumes about our personality and mindset. The posture that we keep our bodies in tells a lot. I have earlier on explained what a straight posture indicates. I am going to add on to the previous information. When observing a person's posture, you should observe whether a person has an open posture or a closed one.

An open posture is when a person keeps the trunk of her body exposed. If you observe it in a person, she is likely to be friendly, willing and open to you. On the other hand, a closed posture is the one in which a person hides the trunk of her body. For example, she will hunch forward or keep her arms crossed. This is the opposite of openness, and the person in question will exhibit hostility and anxiety.

Body Language

Body language is the nonverbal signals that we send through our gestures. In simple words, it is about communication through our bodies. It includes our hand movements and facial expressions to as little things as our pupils. If we observe closely, we will see that people tend to give away a great volume of information through nonverbal signals. As I have already suggested, the key to reading nonverbal signs accurately is to take these signals and study them as a group.

The Eyes

Our eyes are considered as windows to our souls. They are the easiest to learn and most people can do that even without prior training. They tend to reveal a great amount of information about what is running inside our hearts. What we feel or think comes into our eyes. Even a naïve person can take the hint in the eyes of the speaker. But it is not just the eyes that should be studied. Pupils are also very important to know other person's minds. Look out for dilated pupils as they indicate increased cognitive struggle.

Pupils tend to dilate if they are looking at something they appreciate. This is not an easy job to do. If you keep observing different people, you will finally learn how to observe and detect any change in the pupils. If pupils are highly dilated than normal, it means that a person is attracted to someone and is aroused.

Hands, Legs and Arms

Gestures by hands, legs and arms are very important. I'll add on to the previously stated details. Gestures, like our eyes, carry plenty of information about our personalities. Our waving, tapping and pointing have hidden meanings that ought to be understood if you want to master the art of analyzing people. Well, it is important to sort out these gestures as some are cultural traditions like a raised straight palm. In some Asian countries, this suggests hello and in the United States, a thumbs up suggest that everything is fine. You have to keep in view these signs so that they are not mixed up with nonverbal cues.

Coming back to nonverbal cues. If a person has a clenched fist, this indicates anger but in some cases, this also indicates solidarity especially when shown by a politician or a public figure. For a clear analysis, you should study this gesture combining it with facial expressions and speech. Similarly, in some countries, people use the okay gesture that is formed by touching the index finger with the thumb. In some countries, this suggests that everything is going on fine while in parts of Europe, this means that you are nothing. In some Asian and South American countries, this gesture is considered vulgar.

Arms and legs are also quite useful in nonverbal communication. If a person tends to open his arms and keep it that way, he is an attention seeker and full of life. We have learned earlier on that crossed arms suggest closeness and defensiveness. A common gesture that you might have come across is the one in which a person stands with his or her hands on the hips. This is an indication that the person is fully in charge of circumstances and is ready to face anything. In rare cases, it may suggest aggression.

If a person clasps his hands behind the back, he is bored and anxious about something. We have learned earlier on what tapping our feet means. Besides, tapping your fingers also means a lot. It can be a sign of boredom or frustration. When a person crosses his legs, he is closing off on society and wants some personal space. He will prefer privacy than socialization.

Personal Space

More often, we are in need of personal space. Sometimes we want to mix up with people and party but sometimes we need personal space to breathe in. This happens to everyone. You might have been through the phase when you start feeling uncomfortable because of the presence of a particular person. In technical terms, this is known as proxemics. Anthropologist Edward T. Hall explains four levels of proximity between two people. Let's discuss them one by one. (Cherry, 2019)

Intimate distance: Ranging between 6 and 18 inches, this indicates that two people are enjoying a closer relationship. They are comfortable with each other. Two people come at this distance while they are hugging or touching each other.

Personal distance: Ranging between 1.5 to 4 feet, this distance suggests that two people are family members or close friends. If two people keep this distance but are comfortable in their interactions, this suggests how intimate they are in their lives.

Social distance: Ranging between 4 and 12 feet, this physical distance exists between people who have acquaintance with each other. With a coworker, the distance will shorten while with a person whom you don't know well such as a plumber, you will keep it at 10 to 12 feet.

Pubic distance: Ranging between 12 to 25 feet, this physical distance is used in public areas when you are addressing a gathering or a class or giving a presentation to your staff. (Cherry, 2019)

Apart from that, if a person comes closer to you, this suggests that he is looking for a favor from you. On the contrary, if the other person moves away, this means there is a lack of mutual connection between you two. The above-mentioned distance is not something carved in stone. It differs in different cultures.

Mannerism

Winking is a normal act between friends and intimate people, but when a stranger wink at you, it appears invasive and offensive. Wink is generally a break in eye contact which suggests that the person is trying to disrupt the flow of conversation. On a lighter note, while cracking a joke, winking is absolutely fine. Winking without reason, tends to confuse the other person. So, steady eye contact is always the way to go.

If a person has placed his arms in an unnatural position, he is not sure of himself. He is not relaxed and is suffering from a lack of confidence. The conversation with an uncomfortable person tends to be unproductive and inconclusive.

Facial Interpretation

Reading one's facial expressions is an integral part of understanding his nonverbal behavior. We have already discussed some visible expressions like winking, blinking and many other expressions. In this section, I'll briefly discuss micro-expressions. They are brief and involuntary expressions that appear on a person's face. They have great importance because it is pretty hard to fake them. Let's discuss them one by one.

A person's eyebrows will appear to be raised with a slight curve. Their skin just below the brow will appear to be stretched. His forehead will have winkles and his eyelids will remain open for a while. His jaw will appear to be dropping and teeth will be slightly parted. Their mouth will remain normal with no signs of tension.

Pay close heed to a person's lips to detect the element of disgust in their disposition. Look out for if their upper lip appears to be raised or upper teeth appear to be exposed. Also, see if his nose has wrinkles and cheeks, raised. Any such sign shows that the person is feeling disgusted.

You can detect anger from micro facial expressions. She has slightly lowered her eyebrows or drawn them together. Other signs of anger are tension in the lower lid or bulging eyes. In addition to this, if their nostrils are dilated or their lower jaw seems to be jutting out, this also shows that they are in anger.

You also can detect happiness in other people by observing their faces. She is happy if her lips appear to be drawn back. Similarly, if her mouth is parted and teeth are exposed, this is an indication of happiness. Happy people have their cheeks raised eyelids lowered with wrinkles evident underneath. Another common sign is the appearance of crow's feet on the outside of the eyes. An important thing to note is that if she is not engaging her side-eye muscles to show her happiness, her happiness is fake.

Inner Instinct

Inner instinct or gut instinct guides the physical reactions that we give to the world around us. It is the feeling that we sense when our bodies are responding to the processing of information that is stored in our subconscious, as I have briefly explained earlier on. The main purpose of our gut instinct is to give us protection in the wake of unusual circumstances. Sometimes people cannot define it but they are relying on it to deal with worldly matters. Their gut instinct guides them through thick and thin. Its power and influence vary in different people depending on their experiences and spiritual state.

Some people call it a hunch while others label it as an inkling, but in general, it is dubbed as gut instinct or instant instinct. This is different from intuition as it is our primal wisdom, while intuition is our spiritual wisdom. Both humans and animals have gut instincts. In some cases, in animals, this feeling is more powerful than humans.

Take the example of a herd of zebras. Even when they cannot see the lions that are lurking behind the bushes, they somehow sense their presence. When one of them whinnies, the rest of the herd starts racing away for cover. If you are fond of Animal Kingdom documentaries, you might have seen such scenes. Similarly, big animals like elephants rely on their gut feeling to find food and water resources.

If you are a cat lover, you can see that your cat will change its mind once or twice before it jumps over from the second story to the first story. Have you ever heard any story of hikers who got lost in the mountain trails? They had to navigate through the mountains without any compass or anything else to take help from. One of them had a hunch to go to the east and the rest of them followed him. In the end, they had successfully reached the camp. Just imagine what would have happened, had that hiker ignored his hunch.

Sometimes you have a strong feeling that something has happened to your son who is at home. You ditch the office and drive back home to find him unconscious on the floor. If you take a closer look at the world around you and also at your own life, you will find that similar incidents have been happening to you.

Signs of Gut Feeling

There are certain signs to watch out for if you want to follow your gut. The top indication is a sudden feeling of fear, especially if it is uncalled for or totally out of context. The second is a powerful urge to accomplish something just like an inner pull. You might also suffer from chills and shivers in your body. Goosebumps on your arms and body in combination with tingles up your spine also indicate that there is something wrong.

One important thing to consider is that signs of gut feeling differ for different persons. For example, some people may not experience any of the above. Instead, they get nauseous or have physical uneasiness. A few people tend to get alarmed at times while only a handful of people hear instructions or warnings in a clear voice. You might have one sign or all of them.

Discuss Intuitive Cues

Intuition means "to look within." Some scientists term it as sophisticated intelligence. People are viewing it as something that helps us make decisions rather than being a magical thing that cannot be learned. Still, the fact remains that ancient and advanced civilizations like Buddhism, Hinduism and Islam have connected intuition to the human soul. You can see if your intuition is at work by following some simple signs. You will start feeling light and clear in your mind. No emotions will affect your judgment and you will be absolutely calm and relaxed, and even inspired. If you are observing similar signs in your body and brain, your intuition is most likely at work.

Aha Moment

Things come to a standstill at times. A person who is running a clothing factory complains that despite producing the best garments in the market, customers are drying up day by day. He had run a marathon marketing plan to boost up sales but to no avail. Is more marketing the only solution? Shifting the production model can be a viable solution to the problem. Brainstorming new ideas and selling techniques is what we usually do to solve this kind of situation. But what if ideas just stop coming to us? What if nothing seems to be working? Maybe he should freeze for a moment and do nothing. Yes, this works sometimes. He should just stop pursuing a solution to the problem. Instead, he should take a shower, start playing golf or maybe watch a movie. People hit upon amazing solutions to overdue problems when they detach them from the current scenario for a while.

The key to reach the aha moment is creating an environment that is full of silence as well as solitude. These conditions are essential for your brain to nurture these moments. Ultra-quiet places are always the best for making better decisions. Once you have found a quiet place for yourself, you have to start looking inward. Focus on the live stream of thoughts. You have to detach yourself from the outer world like your cell phone and any other thing around you. When the external information ceases to reach your brain, you will slowly start noticing the aha moment. Gradually, you will achieve the "idle" mode of your brain. It is important to know that you don't have to stress out your schedule to get the aha moment. Instead, find a few quiet moments on a daily basis to do this exercise. Also, try to turn off all the electronic gadgets at least for a few hours in a day so that you can leave your brain to wonder for a while.

Human Vibe

The vibe we give off is equally important for reading people as reading their body language. This is closely linked to intuition. We run away from some people and try to be close to some of them. More often, we hear people say that they feel good or bad vibe by being around some people. Some people really elevate our mood when we are around them, while others drain us out of our positive energy.

The impact of the human vibe can be felt when we are just inches or feet away from a person. Some cultures like the Chinese dub this invisible energy as life force, named as chi. Let's take a look at a few examples.

Sometimes, your spouse says sorry to you but you feel that he is not really sorry for his mistake. A coworker is trying to charm you but you know something is fishy out there. A classmate appears to be cheerful but you have already sensed the hidden anxiety. For example, we often say that depression is faceless. People wear a smile in front of others but in reality, they are broken. While most of them around a depressed college fellow ignore her condition, you are sure that she is not healthy at all.

You need to link a person's emotions with his energy to get to know them better. Reading people by human vibe is all about decoding their emotions. By reading people's energy, you can bring yourself in line with how you relate to them, and whether you feel comfortable with them or not. If you study this subject and master it, you are able to make some crucial decisions in an effective way. For example, you will never want to spend your life as a spouse with a person who will drain your energy. The same is the case with a coworker. Why should you consume your time sharing your meals with a coworker who leaves you feeble and unproductive after a single sitting? That's why it is important that you learn how to read the human vibe.

Presence

The first thing to learn while reading people's energy is sensing the presence of people. This is the overall effect that a person leaves on you when he or she is near you. You have to calculate it. A girl in your office may leave mysterious, joyful or sad effects on you. Try to make out if the person around you is pulling you toward her. When you are reading them from their presence, try to notice if the energy they give off is warm or cold. Is it like fresh air or stalled? Do you sense anger or depression when you are near them? Whether it is a friendly sense or an intimate one when she is near you. On the basis of these readings, you can decide how to shape the future course of your relationship with that particular person.

Eye Projection

Another important method to read the human vibe is to take a closer look at that person's eyes. Eyes are the ways to transfer positive and negative energy. In Islamic civilization, eyes are considered as the source to transfer spiritual energy. Sufi poets like Rumi greatly focused on the importance of a glance. They say that the brain transmits electromagnetic signals through eyes. Looking straight into the eyes of your pet releases oxytocin which builds up a trustful and peaceful relationship between you and your pet.

You should take your time when you are observing her eyes, then study what kind of feeling you have. Is it the feeling of love, care, calm or anger? Do her eyes look sexy? Do they intimidate you? People's eyes may feel hypnotic at times. Sometimes looking deeply in their eyes make you feel insecure. That's why you have to study the effects of cautiously. If you come across a negative person, try not to engage them or they will zone in on you. If you sense positivity, keep looking straight into their eyes. Feed on all the positive energy.

Physical Contact

We share our energies with people upon touching them by means of a handshake or a hug. Whenever you touch someone through a handshake, you will know whether the person makes you feel comfortable or not. Or do you just want to withdraw? Do their hands feel clammy? This is a sign of anxiety. They will make you feel anxious. If they hold your hands in a powerful grip that your fingers feel pained, this gives off aggressive energy.

Voice Tone

Last but not least is the tone of voice in which people speak to you. It will speak volumes about their emotions and feelings. The frequency of our sounds creates distinct vibrations. Does their tone soothe you and make you calm? If you observe that the voice tone of a person is so soft that you barely hear him, this shows signs of low self-esteem. If they are too loud, this shows anxiety or insensitivity. If they are fast-talkers in your first meeting, they might want to sell something to you.

Try to observe if people are laughing too much. If this is the case, they are lighthearted. But their laugh ought to be genuine. (Orloff, 2014)

Chapter 5

Lies – Why They Affect the Way You Analyze People?

Lies go undetected more often, so do liars. Lying is quite prevalent among youngsters and this behavior hardly does any harm at that age, but when you grow older and enter professional life, liars can be harmful to your professional and intimate life. Kids consider this habit as something fun to tease their school mates and friends from the neighborhood. When they don't get caught, they consider this behavior as a way to go in life. They integrate this behavior into their personality and use it later on for personal benefits. So, that's how lying as behavior makes its way in our characters.

If you don't nip the evil in the bud, the kids will see this as a baseline for building up a powerful lying pattern to be used in the future. When you are dealing with these grown-up kids, you feel at a disadvantage because they maneuver it so professionally that you realize it only when they have already achieved their goals. It is not that liars are impossible to detect. In fact, they are pretty easy to spot around us. All we need are a few techniques to make out if a person is telling you the truth or is concealing something from you. Before we move on to analyze the techniques, we need to analyze different types of liars to make the process of detecting liars easy and smooth.

Types of Liars

Let's discuss the difference between people who are quite professional at lying. There are certain signs and symptoms that you need to watch out to find out what type of liar you are dealing with. Let's see and analyze each category to gain more insight when you are analyzing people.

Pathological Liar

The first category is the pathological liars. Pathological liars are habitual and they tell a lie in response to any kind of stimuli. They are very good at lying because of the magnitude of practice that they do. They are pretty good at fabricating stories and it is very hard to detect when they are lying and when they are telling the truth. If only you can read their facial expressions and gestures, it is easy to detect them. Look out for the movement of their eyes. If they are trying to avoid direct eye contact, they are not telling you the right thing.

If you want to understand why people lie so casually, you have to understand the circumstances they went through. They adopt pathological lying as a defense mechanism. It is a way to make their way through severe circumstances without hurting themselves. These are not excuses to become a pathological liar but these are the driving factors that push a normal person to integrate this personality trait. By understanding the pushing factors, you will be able to understand why people lie in the first place. In this way, you can stop a pathological liar midway while is weaving his web.

Sociopath

These liars are considered as the worst types of liars. They lie to achieve personal benefits without caring about how it will affect the people around them. They have a heart made of stone and they don't care about other people's emotions and even their lives. In simple words, they feed on lying. Lying is their strategy to get worldly benefits at the cost of the feelings and lives of other people. They don't feel shame or guilt at all.

When you are confronted with these kinds of people, you need to walk cautiously by carefully reading the situation. The situation can go out of control any time and you will find yourself becoming their victim in a snap. The reason is that they can turn out to be amazingly manipulative when dealing with you. They are experts in lying and they are more often quite cunning.

When you are analyzing these kinds of people, you are likely to end up reaching a wrong conclusion because of misleading or insufficient evidence. If you are currently into a relationship with a sociopath liar, you should try to free yourself of the commitment. When you are convinced that the relationship is poisonous because of the lying habit of your partner, end the relationship. You can exhaust your option of changing that person before taking a decisive step. (5 Types of Liars and How to Recognize and Deal with Each, n.d)

By now, you might be thinking that liars are like parasites who drain you of emotions and energy. But we must not forget they too are humans. They are not monsters whose only treatment is to send them in exile out in the wild or kill them with the best weapon available. It is advised that if you detect the lying habit in someone close to you for the first time, you should approach them with kindness. Show them love, tact and affection according to the size and impact of the lie that they just weaved for you or any other person. Don't forget to furnish your evidence or the other person will get away with it by denying it altogether.

It is highly likely that some liars will defend their lies and continue with it when you try to confront them, but you should keep in mind that liars have mastered the art of manipulation. Keep yourself in full senses to get away with their manipulation. (5 Types of Liars and How to Recognize and Deal with Each, n.d)

White Liars

We often see white liars around us. White lies are not real lies. At least they are not as lethal as real lies are. In most cases, they are perfectly harmless, and you can say that white liars more often tell one or the other kind of truth, that's why people believe that they are not lying. Some weak hearted people use white lies in a bid to protect themselves from the truth if they are of the opinion that truth will be damaging or hurtful for them.

When you detect white lies, you should approach those people and try to rectify their ways. If you find out that the white lie has insignificant value, perhaps you ought to let it pass. Otherwise, you can ask the liar to mend his or her ways as it is not a good idea to base a relationship on lies no matter how harmless the lies are. If you fail to detect white lies or let them pass as fun, they may cause serious problems for your intimate relationships in the long run. (5 Types of Liars and How to Recognize and Deal with Each, n.d)

Compulsive Liars

Compulsive liars are habitual when it comes to lying, but unlike pathological liars, you can detect them and figure out how to deal with them quite easily. They are not expert enough to weave a net of truth around their lies to make them appear credible to people. They are easy to analyze because they don't wear a cloak of truth over their woven web of lies. When they speak, you can tell that they are lying because they display such kind of behavior. Things to note when you meet such a person are that they will start sweating, and also they will never look into your eyes while telling a lie.

Compulsive liars can be further categorized into a habitual liar as well as a narcissistic liar. Habitual liars cannot refrain themselves from lying all the time. On the other hand, narcissistic liars make up stories about themselves. They tend to exaggerate things and like to embellish things about themselves. They will you stories how they confronted a dozen warriors and single-handedly defeated them. Other stories include how they turned out to be the hero of a number of situations like saving a girl from a raging fire. Most of the stories they tell may appear to be far-fetched. As per medical science, these kinds of people suffer from a narcissistic personality disorder. Lying becomes their habit because they feel deprived of their real lives. They have reached the conclusion that their real lives are boring and that no one is impressed with them.

How to Deal with Liars

There is a wide range of ways to deal with liars. This can be really difficult but the best approach is not to throw a fit of anger. The liar is likely to channelize your aggression toward diverting you from the subject. The best approach is to avoid getting carried away with their versions of events that you have concrete evidence of not being true. You can deal with liars by being polite and confronting them with the truth.

You have to understand the fact that all of us tell lies at one point or another. Sometimes we have to fabricate a lie to avert a crisis. Sometimes, you need to tell a lie because you don't want to hurt someone's feelings. These kinds of liars are easy to deal with because they tell a lie only to defuse a tense situation. White liars are also harmless unless they make it a habit to tell lies. What if you are confronted with a compulsive liar or a sociopath? They are habitual when it comes to telling lies.

Compulsive liars are not the easiest to deal with. In order to kill their sense of inferiority and inadequacy, they can go on to any extent to tell lies without caring for their effect on the lives of other people. They lack empathy and are unable to understand the extent of emotional turmoil that they bestow on the other people. Their dishonesty takes its toll on others. They are self-centered and can only think about their own benefit and profit.

These types of liars are the most difficult to deal with, but with greater understanding and practice, you can master the art. The first thing you should remember is to avoid confrontation with these kinds of people. They always try not to leave a trace of what they have done. When you confront them, they will come up with a new story to cover up their wrongdoing, in addition, they will become hostile in their attempt to invalidate your evidence. So, there is no point in confronting them. You have to make yourself believe that the person you are dealing with is not normal and he needs help. Think of him as a dysfunctional person who doesn't think normal. If you try to change them, they will resist any effort by hook or crook. So, you need to stop changing them. Just accept them as they are and deal with them as if they are normal. This will make them friendly toward you and it will be easier to deal with them.

The next step is to listen to what they say carefully. Don't trust it right away. It is better to retain the factor of doubt. Spare the room for verification of what they tell you. You should be careful about not letting them know what you are up to. When you are sure that the person is a compulsive liar, put a limit on the time that you spend with him; otherwise, they will keep draining you of energy and demoralize you.

These types of people don't merit your time and love. Avoid sharing your personal information and any other details with that person. Don't open up too much or they will use that information for their personal benefit. They can do that without thinking even once because of the fact that they don't have empathy for others. (Kloppers, n.d)

The next step you should take care of is that you must not expose a liar. You think that you have detected a liar and the liar also knows that. As per your impulse, you will rush toward your closest friends to tell them about that person in order to save them from his or her heinous behavior. Freeze and think for a moment. Is it really a good idea to tell others about him or her nature? The answer is a 'no.' In fact, it is pretty dangerous. The liar will behave like a suspect does on getting detected by cops. Move on in your lives as if nothing happened. Focus on what you are doing and you will be in complete comfort. In some cases, if the liar is bent on inflicting losses and pain on you, you have to do something about it. Even in these situations, think about the possible impacts of exposing them. Better have a comprehensive discussion on the subject with the people who are close to you.

When you have decided to expose a liar, you should do this carefully so that you don't paint him or her in a negative picture. Try to convince others that he or she did that out of sheer necessity. This will paint your picture positively in the eyes of the liar. That's how you can expose him and also succeed in gaining his sympathy. In fact, this can promote friendship between you two.

In severe circumstances in which the chances of confrontation are high and moving on also is not a good choice, the only way is to show that you understand why the liar committed that wrong act. Not only show them by gestures or expressions but also try to tell them in clear words that you understand why they did that. Tell that it is normal to do that for self-protection, and also tell them that you accept them. That's how we are actually telling them that what they did was wrong but we are forgiving them. This has the potential to change their hearts. Perhaps they decide to mend their ways.

The above method doesn't always work. Some people tell harmful lies without shame or regret. They even inflict serious losses on people by telling lies. They are the ones who ought to be exposed so that other people should be saved from their lethal actions. You must not fear of exposing them and getting into a direct confrontation.

They have already inflicted losses on so many people that there will be hardly anyone left who will show sympathy toward them. The people whom he has done wrong will support you. When we are done exposing them, we should immediately part our ways with them and become more cautious.

Compulsive liars are without a doubt hard to deal with. There is no hard and fast rule for the purpose. You have to read the person and then tailor your reaction to suit the circumstances. Without much homework, you will only land in trouble.

Chapter 6

Adverse Effects of Misreading People

This chapter is going to walk you through the effects of mixed signals. You will learn how people misread each other's signals and how it lands them in trouble or at least create a web of confusion among them. Reading people, though seems easy, is a tough nut to crack. If you miss out on a key signal and misinterpret it, you are going to misread someone's intentions. Bad intentions will be interpreted as good while good as bad. Similarly, wrong judgment will hamper our social connections or relationship with our colleagues.

This suggests that you should know the consequences of misreading people so that you may remain cautious. One wrong decision may land you in trouble. Mixed signals are dangerous in the sense that they confuse you, and confusion edges you off the right track. You should know what mixed signals are and how you can deal with them to avoid a crisis situation. This chapter also carries examples of mixed signals and the ways to tackle them wisely.

Mixed Signals

I have a friend named John who got a job at a grocery store. There he had a team of around a dozen people. As a good manager, he used to call a meeting every Thursday. Each meeting had an agenda that John followed in letter and spirit. John tells me that he wanted to be as handy to his staff as was possible. He used to help them in packing and putting groceries on the shelves for display.

John believed that he couldn't do more for the staff. Unlike most other bosses, John was a really good listener. He always welcomed criticism, suggestions and new ideas to improve the look of the store and boost sales. He was pretty satisfied with his role in the store. One day he welcomed criticism on his own performance in the store, so he requested his staff to criticize his shortcomings. Literally no one appeared at the meeting, suggesting that they had no issues with his style of running the store. After insisting for a week, one employee appeared in his office and opened up his heart. He said, "Mr. John, why do we think that we cannot do our jobs right. Why do you always come up to give us a hand? Don't you have confidence in our abilities?" This was completely shocking for John. He didn't think that his offer to help his staff would be perceived in another way.

John thought that by helping her staff, he would be able to identify with them. They will feel relaxed and satisfied, but the result was completely different from what he had thought. Instead of considering John a generous and kind person, his team felt a kind of inferiority complex.

This kind of scenario may happen and we don't even know about it. John's staff misunderstood his intentions. They perceived it as a lack of faith in their ability to do the job efficiently. These misunderstandings herald conflict as well as resentment, and this kind of misunderstanding is pretty common between couples. This story is also related to John. Compelled by his kind and affectionate nature, John wanted to hire a maid for her wife who was pregnant, and he did that accordingly. On the contrary, his wife thought that John didn't like the food she used to cook for the family just because he had passed critical comments on one or two dishes she had cooked. Despite the fact that you have explained that you are hiring the maid to relieve your wife of workload, but the seed of misunderstanding has already been sown.

Take a critical look at your own married life. There might be more than one occasion when you and your wife misunderstand each other for insignificant reasons. For example, you might be dining while your wife is telling a story to you. Although you are all ears to your wife yet she might find the act of your dining while she is speaking offensive. This may lead to a potential misunderstanding between you two. Similarly, you two have made a plan to go to a beach for sunbathing. Your wife feels sick and excuses herself from going with you. Although her excuse is genuine yet you might think that she doesn't want to go with you. Similar incidents of misunderstanding may happen when you two disagree on simple things like watching a movie together.

So, there is usually a big gap between what we say and how our listeners perceive it. The difference between real meanings and perceptions is not always a matter of egocentrism. Mixed signals are complex to understand but they have great importance when it comes to reading people. Mixed signals confuse you and land you in a blind spot where you cannot think clearly or see things how they are in reality. These signals cloud one's judgment of people and circumstances.

Take the example of a dating scene. You are dating someone who is not responding to your texts, but after some time, she reads your Whatsapp status or Facebook story. You will be confused. The point to understand is that we, as humans, lack perfection when it comes to expressing our thoughts. This is also true that we improve on our experiences and try to streamline our understanding of others' thoughts. Still, our true feelings tend to get hidden in how we translate them into actions and how we communicate through our speech. So, we can say that mixed signals are negative signals because that's how we are going to perceive them. One in hundred, if not thousands, will see something good in a mixed-signal.

So, should it go that way? Is it destined to be that way? There is an antidote to this problem. When you are confused about the inner feelings of a person, you should read their words coupled with their actions. But this demands practice, a lot of it, to decipher that hidden meaning accurately and perfectly.

Why Do People Give off Mixed Signals?

If you are receiving mixed signals, all burden is not upon you for reading her accurately. She also has to streamline lots of things. Mixed signals lead to miscommunication in most cases, and this affects the health of your relationship. Sometimes people intentionally use them to keep someone at arm's length because they just don't want to engage with them. For example, your fiancé is fed up with the relationship, but she cannot express it in words as it would be hard to hear and also, it would lead to an endless debate which she definitely wants to avoid. Here, she will start sending mixed signals to you. Like ignoring your texts but talking to you on phone or ignoring your call and responding to your texts, so that you take the hint that she doesn't want to be with you anymore. That's why first she will slow down the pace of the relationship and then she will say goodbye so that everything concludes making the least possible noise.

The story doesn't end here. Mixed signals don't always mean that the other person is trying to avoid you. It also is a way to cope with the stress that comes from getting intimated and close to other people. Your girlfriend might be going through this phase of stress, and you unknowingly end the relationship blaming her for intentionally avoiding you. Let's take a look at some mixed signals that sabotage relationships.

They Don't Meet up Your Expectations

There might come a moment in your life when you keep waiting on end for a special person in your life to respond to your texts or Whatsapp status. It is normal behavior to send and text and then expects a response to it right away. Absence of which can cause confusion and misunderstanding, and may mar your relationship in the long run. It is normal that the other person might be caught up in work. You will wait for the first few minutes but when a considerable length of time has passed, frustration will come to hit you hard. You will start feeling off about it.

It is possible that they will respond to you when they are free and when they find it convenient. In order to have a clear view of the circumstances, you should note if this kind of behavior has become a habit with them or not. One thing is clear from a recurrent behavior that the person is not fully dedicated to you.

Half-hearted Effort to Meet You

"I am dying to meet you. When will we meet? I am planning to drop in this weekend. Stay free." She texts you thrice a week but has not yet found time to come over on weekends and spend time with you. Every time she misses a weekend, she texts you saying that she remained busy. One or the other assignments keep her from coming over to you. She says that she has to juggle responsibilities and priorities. You remind her that she is placing other things as top priorities and ignoring you. People are not busy at all. It is all about priorities. When she has decided to meet you, she will find a way out. If she is not doing that, she has other things at top priorities. That's why she is unable to fulfill her commitment to you. Maybe she is sending you a mixed signal for a reason. Take the catch and make a decision.

She Doesn't Open up as She Should Be

When a relationship kicks off, you expect your partner to share everything with you like the names of her friends, information about her exes and lots of other things. It is this transparency that helps in cementing the foundation of your relationship. When the two of you have shared everything with each other, you will be able to form an emotional connection, which sets things off. Both you and your partner need to share their bit for a healthy connection. If you are sharing everything while she seems to be holding back, this is not a good omen for the relationship. The foundation will have cracks right from the start, which will eventually bring down the entire structure one day. Therefore, if you sense such a behavior, take it as a deliberately mixed signal. Analyze it and make a timely decision instead of delaying and regretting afterward.

Does Your Partner Flirt with Other People?

This turns out to be painful than other signals for lots of people, but this is also an important one to study if you want to make accurate assessments. This happens in thriller movies. The hero has a girlfriend who is a bit friendlier with his friends. At first, everything seems to be normal but slowly, you realize that she is up to something else. Do you remember a scene from any Hollywood movie in which hero plans for camping beside a lake along with his girlfriend and college fellows? At the campsite, she pays more attention to the friends of the hero. The hero gets confused at first due to the mixed signal. Then gradually, he realizes that something is fishy. His girlfriend is not actually interested in him. She is keeping all options open so that if one doesn't work out well, she could jump to the other.

The solution is not to frame allegations around her but keep patience. Ponder over how she is dealing with your friends. Note the dialogues, the gestures and the frequency of their meetings in addition to the time she spends with them. When you are sure that something is wrong, you should take her out for a walk at someplace where your friends couldn't reach to disturb you. Now you ask her in clear words about what is happening and why is it happening? You can request her to change her behavior because the current behavior unsettles you. If she truly cares for you, she will try to tone down her behavior and keep herself in check. If she doesn't try at all, take this mixed signal as a clear sign to make a decision. It is better to part ways than regret afterward.

She Cares for You When You Are Alone but Doesn't Show Affection When You Two Are in Public

Watch out for this mixed-signal carefully. She is ready to make out with you while you are at home. She is super comfortable while talking to you on end and watching a movie with you, but when you are out with friends for a hangout, she is unwilling to be seen with you. She just doesn't want to open up about her relationship with you. If your relationship is in its infancy, you should give your partner some time to adjust herself in your and her friend circles. When she is comfortable enough, the relationship will take a smooth road and move on well, but if she continues to behave like that after a while, you should take this signal with caution. Perhaps she made a hasty decision and now she is regretting. She may be pointing toward the underlying tension that exists in your relationship. Maybe she doesn't want to be seen with you anymore in public but is too polite to tell you so.

Remember that when a person truly loves you, their words and actions go on well. If she promises you to show up at your office when your boss throws a party for you and she doesn't keep it, these signals should be taken as serious. Like all other mixed signals, you have to give her some time. After three to four incidents, you will be better positioned to make a decision.

Jason Miller

Chapter 7
Analyzing Verbal Cues

In order to know the difference between the truth and deception, you have to follow certain cues. The signs of lying are not clear; hence they are hard to understand. In addition, you cannot always be sure whether a person is lying or not. By practice, you can be able to tell if someone is lying to you or not easily. The rule is simple. When we are lying, we are deviating from how we behave naturally. We have to make an effort to look truthful while we are lying; that's why if you know how a person behaves naturally, you can easily tell when they are lying by tracing the difference in their behavior. The difference can be the inclusion of certain words or phrases that he normally doesn't use.

Look for Deviations in Their Words

Inconsistencies can help you distinguish the truth from the lie. For example, a person at your office tries to convince you that he didn't meddle in your documents. If he is telling the truth, he will not care what you like to listen to from him. Otherwise, he will formulate a plan in his brain. He will brainstorm what words and phrases should be used so that he may look truthful before you.

The phrases like, "I didn't do that. I wasn't in the office at that time. How could if do it?" should be enough to put him under suspicion. In addition, he will repeat these words and phrases again and again. Experts believe that this kind of repetition buys them more time to think and fabricate another phrase that could convince you that he didn't do it.

Another verbal cue is that he will tell you more than you need to listen. Chances are high that he is telling you a lie. Liars talk too much because they have made it a habit to fabricate lies. They're uncalled for openness should be enough to alarm you.

Another indication of a liar is that they find it pretty hard to speak when you try to ask questions from them. They will stammer, lose words and find them entirely speechless. The reason for this kind of behavior is psychological. Their mind is not ready for rapid questions. Liars make up stories when you ask them a question. After one or two questions, they find them at a loss. Another reason is that our automatic nervous system malfunctions during stressful times. This dries them out of answers, which is an indication that they are telling lies. Also, watch out if they are biting or pursing their lips or not. Any such behavior is an indication of a liar.

Learn to Ask Right Questions

Parents have to believe what their kids say to them. When they say they were with their best friends whom you know very well, you believe them without investigating the truth. But when they tell the same thing again and again, this means there is something fishy in the bottom of the story. Teenagers want to do lots of things that pass their mind and to make it possible, they tend to tell lies to their parents so that you their parents or teachers approve of their activities. When they suspect that a particular activity would not be approved, they tell outright lies. This is the time to worry.

If you level allegation of lying against them, they will become hostile to you right away, and this will only make them more stubborn. That's why you need to be tactful to make them realize that their lie is not working without them knowing that you are manipulating them. That's where you need to use the technique of Volatile Conundrum. Try to create a scene. Ask your son the right question. Ask him where he went with Jimmy, the name he used to deceive you. Jimmy is his classmate whom you approve of if your son remains with him.

"You got home pretty late at night."

"Where did you go with Jimmy?"

He would say that they were at McDonald's to celebrate the birthday of their friend from school.

Here you have to come up with your own version of the story. "Really? I heard that a minute fire broke out at McDonald's due to short-circuiting. Did everything go well? When did the fire brigade reach the site?"

Now, this is the momentum where your son will be caught in a conflict. Whether to approve your version of the story or deny it altogether? If he approves of it right away in a snap decision, you have successfully caught a liar without confronting him. If he disputes the fact that the fire didn't break out but in reality it did, again you have successfully caught a liar. In this way, you have successfully put your kid in a Volatile Conundrum situation.

Knowing How and When to Read Verbal Cues

All of us use verbal cues almost every day. Have you ever wondered how do you communicate with people? What are the ways in which you communicate with them? Communication is not a simple process that you can easily understand. It is rather a complicated process that is so detailed that you cannot miss out on a single nuance without miscommunicating what you have on your brain. There are little things that you take into account during communication such as your reaction when someone tells you a joke. Whether you should laugh, smile, or don't do anything at all. We usually get ready to laugh when we are sure that the person has delivered her punch line of the joke. Some laughs are spontaneous. You just cannot wait to understand before you laugh whether it is the punch line on which you are laughing or not. So, that's complicated. What if you laugh before the punch line, would it not sound awkward? What if you have delayed the laugh? Now the other person will be in an awkward situation.

You have to look for verbal cues when you are communicating with someone. In communication, cues are generally considered as prompts that you can use to show others that it is time for them to issue a response or give a reaction. A verbal cue can be a word, a pause in language, rise in the tone or fall in it, or anything else related to speech. For example, I asked my friend, "Shall we try our luck in starting a new business for the two of us?" Now I have put up a question for my friend and I expect a response from him. There should be an answer or the communication will hit a stumbling block.

Verbal cues are more important when we have to teach children at home or at school. Children are not so accustomed to understanding non-verbal gestures like facial expressions and body language. You have to explain everything in words before them. When a teacher has taught kids a lesson on the whiteboard. She asks them, "Can anyone draw a circle on her page like the one that is drawn on the whiteboard?" She will for sure use nonverbal gestures like pointing her hands toward the circle and toward the pages that are put in front of them on the desks. So, that's how with the help of clear words teachers are able to communicate their questions and instructions to the students.

Take another example. The teacher has taught the kids about circles and the way to draw them. They come the next day to the class. The teacher plans to take a surprise test about circles. She will draft a question in her head that will be easier for the kids to grasp and respond to. Perhaps she says to them, "You remember what we learned yesterday?" At least a few of them will respond in the affirmation. Now she says, "Who will come up and draw a circle on the whiteboard?" This is the question that the kids will understand and respond to you accordingly.

Direct and Indirect Verbal Cues

You need to know the words when communicating with other people. Direct verbal cues are clear statements or instructions. Parents are quite skillful in these verbal cues because they have to raise kids. Even new parents find out ways to train the children because verbal cues are integrated into our nature. Let's see some example of verbal cues that a child understands easily and integrate into his or her brain to use it in the future.

- Come to me.
- Go and clean your bedroom.
- What are you chewing?
- What are you studying?
- Why have you come so late from school?
- What are you thinking about?
- Have you brushed your teeth?
- Did you put the blender on the rack?
- Where are your books?
- How did your exam go last week?

So, these are the questions that we ask our kids every day. These examples contain clear instructions for the kids that's why they understand them right away and respond accordingly.

The second type of cues is indirect verbal cues. These also are considered as prompts but they are quite less obvious than the direct cues. I mean they are just not direct questions with a clear question mark at the end. When a teacher shows up in the class and puts the following questions?

- Has anyone seen my pen?
- Has anyone got an electric clock?
- Have you understood the concept well?
- Does anyone know how to draw a circle?
- Will anyone show up at the desk to draw a circle?

These questions are not specific to a single student. Instead, these are general questions. Only the students who will relate to them will respond to them accordingly. In simple words, we can say that indirect verbal cues throw the ball in the court of the listener. It is him or her who will decide whether to respond or not, how to respond and when to respond. The prompt in indirect verbal cues are not directed to any specific person. See the following examples:

- What are you going to eat today?
- What have you done from dusk till dawn today?
- What work have you done to clean the house?
- How did you bake the cake?
- How are you going to get a job in NYC?
- What are you going to do in the evening?

Chapter 8: Looking into One's Own Self

It is a proven fact that magic crystals, tarot cards, palmistry and astrology can help develop your psychic skills but still, the most direct and effective method to know about yourself is to connect with your own mind. If you really want to connect with your own self, you will have to invest considerable time in reading your habits and how you behave. Just like meditation, you have to stay away from television, radio, and mobile any other activity that would engage you to mind. There should be no children or pets around you while you are on your way to finding yourself. You can turn on light music if it helps collect your thoughts but you can also sit in complete silence if it makes you comfortable. Let's take a look at some key benefits of self-knowledge.

Benefits of Self-Knowledge

There are certain benefits that you need to take a look at in order to be motivated for exploring yourself.

- Knowing yourself will offer you a special kind of pleasure and happiness. You are in a position to tell other people who are. Your expression is confident and smooth. When you know what you desire for, you can express it in simple words.

- Knowing yourself helps you improve your decision-making. When you tend to know yourself, you are better able to make certain choices about the world. These can span from making small decisions to big ones like choosing your partner. You are more ready to tackle the problems of your life and also find solutions for them.

- Knowing yourself offers you self-control. The ability to know yourself helps you understand what motivates you to put a stopper to bad habits and what is needed to adopt good habits.

- Good knowledge of your own self helps you resist social pressures that are constantly mounting upon you from one or the other sides. When you know what you like and dislike, you are more ready to say yes and no to certain people and their proposals.

- This also makes you more willing to tolerate and understand other people. You are in a good position to know your own struggles which helps you identify with other people. This instills more tolerance in your personality. (Selig, 2016)

Let's see how you can know yourself.

Concentrate on Yourself

Before you go on to knowing yourself, you should clear your mind first of any intrusive or lingering thoughts that come to obstruct your mind. Bring yourself in a position in which you are the least distracted. Just focus on the current moment. You can try to focus on an imaginary point in your brain. Stabilize that point and try to find a grounding place where you find harmony. You need to focus on that point until your brain is free of negative thoughts. Concentrate on the white light of your consciousness. Feel the calm this state has brought to you. When you are no more distracted by negative thoughts, you can move on to the next step.

Ask Questions

Throw questions before your psychic self. This is where you can start thinking about your life and get answers from your own self. Before going into this procedure, you need to have a clear idea of what you have been trying to find out about you. It is always a better idea to jot down these questions on a piece of paper and memorize them. Now ask them from yourself. See the following examples:
- What is the perfect job for you?
- Where do you want to live?

- What type of partner do you want to have for you?

Try to be as clear as possible in asking questions. Vague questions will only produce muddled answers.

If you are doing it for the first time, it will be hard to get answers in the first go, so if your brain is empty of answers, don't take it to heart. Instead, keep trying to explore yourself. Give yourself time and space to settle on what you are trying to ask it. Keep your body and emotions in a fair check. You might feel unexplained sensations in your body or some emotional reaction. Don't ignore them. Note them down and try to see what they are trying to explain.

Gradually, you will be able to find the much-needed answers to your questions. Persistence and the right practice are keys to it. If you start curbing your emotions, you are binding your brain which is not good. Let every emotion and feeling flow naturally so that they may aid you in finding the right answers.

Know Your Personality

You should have complete knowledge of your personality. You think that you know yourself because you know who you want to meet, what you like in food and what you dislike, how you want your partner to be and behave. But have you ever experienced a situation in which you couldn't explain how you reacted in a certain way? We deal with certain people and things which we regret later on and even feel ashamed of. Still, we cannot explain why we reacted that way. How do you react to failure, success, a challenge or a bad day? All these things matter much.

Find out Your Core Values

Your core values, moral codes, and principles always remain dear and near to your heart. There are certain values on which you just cannot compromise. These values will ultimately affect your decision-making ability, the power to resolve conflicts, your way of communication and your day-to-day living style. Find out what they are by deep introspection, as I stated in details at the start of the chapter. Are they honesty, flexibility, integrity or security? Are you soft-hearted, dedicated to the cause of others, prone to learning, wise or a leader? Once you have agreed on what your core values, you be more than ready to analyze other people and also mend your own ways when you stray away from the right path. (be your own psychic – 5 steps to give yourself a psychic reading, n.d)

Know Your Body

Our body is as complex as our brain is. Whenever you start to know it, it changes. When we are children, it is pretty different than when we get old. It remains a piece of a mystery until death because we don't take an interest in exploring its limits. It is full of surprises. Sometimes these surprises are positive while at other times, they are absolutely shocking. Did you ever think what your breathing pattern is? What are your abilities? How flexible are you? How much balance can you bring in your walking pattern? (be your own psychic – 5 steps to give yourself a psychic reading, n.d)

There are times when we say no because our body has reached a certain limit. I cannot do this or I cannot do that. Our body feels challenged. Here you need to take the time to become intimate with your own body such as your strengths and weaknesses. Whether you are comfortable in cold weather or hot weather or balmy weather are things that you must know about you. Many people claim that they know themselves but in reality, they are missing out on clarity. They are just not clear about their mind and vision. (be your own psychic – 5 steps to give yourself a psychic reading, n.d)

You Need to Know Your Dreams

All of us have dreams of a great work future, kids and a luxurious lifestyle. We dream about so many things that we get confused which is the thing that we want more. What are our preferences? Knowing dreams are important and they are worth going after. Get to know them and prioritize them in your brain so that when someone asks you, you are able to speak about them clearly without stammering or repeating.

If you are confused about a dream, ask yourself if you want to do a certain thing. For example, you want to become an interior designer. Gather all the details about this profession. Now ask yourself if you can accept this profession with all its intricacies and liabilities. If you find the answer in affirmative, you need to integrate your dream in your daily pursuit of goals. If you find out that the dream existed in your mind without any reason and that you are not sure whether to pursue it or not, just discard it and never let it distract you in the future. (be your own psychic – 5 steps to give yourself a psychic reading, n.d)

Know What You Like

We believe that we know what we like but in reality this is not true. When someone knows himself, he is highly confident when dealing with others and doing some kind of work. The confidence is evident in his acts and speech. Almost every one of us gets carried away with the popularity of things thinking that we like them but the feeling wears away with time, leaving you confused.

Knowing yourself means that you know your likes and dislikes up to the extent that you are able to write them down on a piece of paper without thinking much when you are asked to do that. Ask yourself the following questions. (be your own psychic – 5 steps to give yourself a psychic reading, n.d)

- What are the foods that you like the most?
- Who are the people you like to meet more often or who give you a pleasurable feeling?
- Which fruits do you love to eat?
- Which vegetables are your favorite?
- Which family members make you feel comfortable when they come to meet you?
- Which friends are annoying to meet?
- Do you like mobile games?
- What type of clothes do you want to wear?

You need to start learning by looking into the mirror. Sort out what you like and what you don't. Now all you have to do is to stay true to your likes or dislikes. If you keep doing what you don't like and also ignore what brings you joy, you are doing great injustice to yourself.

In fact, you have become ready to give up your own personality. In simple words, you are not going to be happy. On the other, hand, if you take care of your likes and dislikes, you are more likely to be happy. (be your own psychic – 5 steps to give yourself a psychic reading, n.d)

Practice makes you perfect. The more you practice, the better you will get on reading people. When you know yourself, you are better able to see others in a clear

Conclusion

Social cognition means how we understand people. This enables us to predict how they will behave and how they will share certain experiences. In addition, it is also critical that we understand certain nuances in everyday speech to make out the hidden verbal cues in the speeches of our colleagues and bosses. Many a time people don't mean what they say and don't say what they actually mean. For example, when someone says, "it is getting cold." It indirectly means that you should go and close the window or the door. You can easily understand the hidden meaning in the remark.

Practice makes us understand what is running in the minds of people, even if they don't speak it out. This is how we can understand their beliefs, experiences and feelings. When we place ourselves in others' shoes, we tend to learn how they think and will behave in a certain situation. This is the start of our understanding of our colleagues and family members.

Reading people is complex. Have you ever had a look at a person and figured out how that person thought or what his nature was? Did you reach the right conclusion? Or did you make a mistake right from the start? The conclusion doesn't matter. What matters is that you tried to make a judgment. If you are always right about your judgment, you are a lucky person because there are so many people in the world who have to go through lots of reading and practice sessions to be able to read other people perfectly. You always need this skill, whether you are an executive in a company who has to run a team of a hundred people or an employee who has to do lots of work and keep his boss happy. The need for reading people increases when you change a job or meet a new boss. Only after a careful judgment of that person, you are able to better communicate with them.

Similarly, at home, you have to read the mood of your father and mother, especially when you have to communicate something important such as your marriage proposal, some girl or boy you like or about the future of work. Only when they are in a good mood, you can be able to say and be heard positively what you want to say. Perhaps you have scratched their favorite car so you will have to catch them in the right mood to communicate that tragic news to them. If you misread them, you will land yourself in great trouble.

Reading people is important and there is more than one reason for that to prove that this is a good skill to add to your skillset. Now that you have gone through the book, you can understand that reading people is essential before you approach a person to talk to him or her. If that person looks friendly, you can go on and open your heart to him; otherwise, you may decide to hold your feelings a bit longer. This skill can enable you to judge if your friend is upset. You can go on to know the reason of his disturbance and help him accordingly. If you are a master of reading and analyzing people, you are very well on the road to success at your workplace. You have to meet people who have different types of behavior. If you misjudge a cunning person and tell him your secrets, you have brought doom to your life by your hands. Similarly, if you have misjudged a sincere person and kept him at bay, you are missing out on a pure friendship that could have helped you climb the ladder at your workplace.

In addition, if you have to gain expertise in the skill of reading people, you are well on your way to be a social magnet. You can easily read people and judge the situation and tailor your communication accordingly. That's how you can win lots of friends and get popular in your social circles. For example, if people appear to be friendlier, you can approach them with a smile on your face and informal greetings. Otherwise, you can take up a formal persona and deal with them accordingly. So, reading people helps you take up a fluid personality that you can shape up according to the expectations of those around you.

This is a general rule. When you say things that others want to hear or behave as others expect you to do, you become a popular figure in your circles because you have mastered the art of keeping them in comfort zones those near you. Your social circles will remain full to the brim always. Understanding the feelings of others is an art that helps you anticipate what is running in their minds, which can help you tailor your speech.

The world is full of confusion. Misreading people leads to a flawed judgment that in turn leads to an inaccurate assessment. Sometimes, a misread facial expression can lead to cracks in the relationship and cause the death of it. For example, she loves you but just doesn't want to talk to you because she had a bad day at work, but you misread her facial expression and distance yourself from her. No matter how nicely she explains her position to you until you read it yourself, the element of doubt will remain in your mind. This small element can plague the entire relationship in the days to come.

So, reading people plays a crucial role in shaping up your intimate relationship. In addition, it can help you at your workplace. This book has walked you through the methods you can use to read people. These methods include reading people with the help of understanding their body language like the movement of their hands and arms, how they sit or how they walk. I have also explained in the book how you can read the facial expressions of a person to judge what he is thinking or what he has to say to you. You can read people by some pretty micro facial expressions to better your judgment of them.

A chapter in the book explained different types of people and different personalities that people take up to move through this life so that you have a better know-how of which type of personalities exist and what is the mindset that is linked to each personality. This will help you better judge people when you are able to identify them with a personality that you have read and integrated into your brain; the process of reading them gets smooth and easy. You also learned about the gut feeling and how it plays a crucial role in guiding your decision-making in day-to-day activities. In addition, you learned about the human vibe and how it can be linked to reading people. How you can study emotional energy and know how the other person makes you feel when he is close to you and how is he going to deal with you and whether you should keep in contact with him or not for the long term.

The book also explained different types of liars like what are their types and how they are you can deal with them. What steps you should avoid and what steps you must take to tackle them. You have learned the adverse effects of mixed signals if you misread them. Mixed signals have ruined lots of relationships and it continues to do so, just because we lack skills in sorting them out, and we always make a hasty decision.

I hope you have learned a lot and have started sorting out things in your brain. We have the basics of reading people in our subconscious. All we need is to sort it out by studying what a typical reaction means and then start implementing it on our social interactions.

References

be your own psychic – 5 steps to give yourself a psychic reading. (n.d). Retrieved from https://www.micheleknight.com/articles/psychic/psychic-ability/be-your-own-psychic-5-steps-to-give-yourself-a-psychic-reading/

Cherry, K. (2019). Understanding Body Language and Facial Expressions. Retrieved from https://www.verywellmind.com/understand-body-language-and-facial-expressions-4147228

Chu, M. (2017). The Truth About How Gut Instincts Really Work. Retrieved from https://medium.com/the-mission/the-truth-about-how-gut-instincts-really-work-d665425f1eb1

English, J. (2019). 5 Basic Body Language Signals of Manipulators. Retrieved from https://drwebercoaching.com/5-basic-body-language-signals-of-manipulators/

How To Read People Like the FBI. (2018). Retrieved from https://www.thrivetalk.com/how-to-read-people/

Kloppers. M (n.d). Dealing with Liars. Retrieved from https://www.mentalhelp.net/blogs/dealing-with-liars/

Orloff, J. (2014). The Power of Surrender: Let Go and Energize Your Relationships, Success, and Well-Being [pdf]. Retrieved from https://www.amazon.com/Power-

Surrender-Energize-Relationships-Well-Being/dp/0307338215/ref=as_li_ss_tl?ie=UTF8&redirect=true&linkCode=sl1&tag=theminwor01-20&linkId=7bec015a8cfbec80e5bb69f63a7ca784

Scott, R. (n.d). How to Read Body Language – Revealing the Secrets Behind Common Nonverbal Cues. Retrieved from https://fremont.edu/how-to-read-body-language-revealing-the-secrets-behind-common-nonverbal-cues/

Selig, M. (2016). Know Yourself? 6 Specific Ways to Know Who You Are. Retrieved from https://www.psychologytoday.com/us/blog/changepower/201603/know-yourself-6-specific-ways-know-who-you-are

9 Personality Types – Enneagram Numbers. (n.d). Retrieved from https://www.theworldcounts.com/life/potentials/9-personality-types-enneagram-numbers

www.ingramcontent.com/pod-product-compliance
Lightning Source LLC
Chambersburg PA
CBHW072003070526
44583CB00015B/1309